D1266981

Contemporary animal learning theory

Contemporary animal learning theory

Anthony Dickinson
The Psychological Laboratory
University of Cambridge

CAMBRIDGE UNIVERSITY PRESS
Cambridge
London New York New Rochelle
Melbourne Sydney

Published by the Press Syndicate of the University of Cambridge
The Pitt Building, Trumpington Street, Cambridge CB2 1RP
32 East 57th Street, New York, NY 10022, USA
296 Beaconsfield Parade, Middle Park, Melbourne 3206, Australia

First published 1980

Text set in 10/12pt Linotron 202 Times, printed and bound in Great Britain at The Pitman Press, Bath

British Library Cataloguing in Publication Data

Dickinson, Anthony
Contemporary animal learning theory. –
(Problems in behavioural science).
1. Learning in animals
I. Title II. Series
156′.3′15 QL785 80–40764

ISBN 0 521 23469 7
ISBN 0 521 29962 4 Pbk

Contents

Foreword

The space in which science lives and grows has some curious features. It is natural in some ways to imagine a tree diagram: a thick trunk (philosophy?) divides first into a few major branches (physico-chemical, biological, social sciences), then into somewhat thinner branches (physics, for example, now separating from chemistry), and so on repeatedly until the twigs that display the latest, developing buds ramify in all their recondite glory. The trouble with this image is that it leaves no room for the fusions between long-separate branches which make up one of the most powerful driving forces in contemporary science. If biology and chemistry once split off from each other as branches that aim at different places in the sun, how can we put them together again in our tree diagram to symbolize that modern prodigy, biochemistry?

Another image, avoiding this problem, is that of the bicycle wheel: a central hub (philosophy?) with spokes radiating in all directions, the distance between spokes being proportional to the distance between the sciences they stand for. Now, to celebrate the birth of biochemistry, we merely have to slot in another spoke, equidistant between the parent disiciplines. And, to preserve the necessary metaphor of growth that is integral to the tree diagram, we can make our spokes expand proportionally to the maturity of the subjects they represent. But it is a consequence of this image that spokes get further and further from each other as they grow: the quickest route between two disciplines is always through the philosophical hub; and, as subjects mature, they become more isolated, more idiosyncratic. To be sure, this feature of 'bicycle wheel' space is, to some extent, veridical: Bear witness the latest journal devoted entirely (I invent, but barely) to the eating habits of the *ob/ob* mouse after lesions to the ventromedial nucleus of the hypothalamus. But, from another point of view, it is the growing points of science that are the closest together, the circumference of the wheel that is shorter than the hub. To take an example close to our interests, it is those workers who push the frontiers of brain science forward most vigorously whose research may be transformed overnight by developments in physics, chemistry or linguistics; the linguists, whose subject may be revolutionized by advances in brain science.

The bicycle wheel is ill-designed for these paradoxes. Our series, *Problems in the Behavioural Sciences*, in contrast, is designed precisely to house them. Its intention is to provide a forum in which research workers on any frontier that relates to psychology *de facto*, even if not *de jure*, can communicate their discoveries, their questions and their problems to their peers in other disciplines, and to a new generation of students in their own. The central role played by psychology in this series guarantees that it will be multi-disciplinary; for psychology already draws on advances in fields as diverse as biophysics and biochemistry, logic and linguistics, social anthropology and sociology. Conversely, the increasing use of behavioural methods – often, alas, ill-understood and worse-applied – by scientists in other fields of biological or sociological enquiry means that they too need to keep a wary eye on how the battle rages at the frontiers of psychology. They will find in *Problems in the Behavioural Sciences* the latest war reports.

Contemporary Animal Learning Theory is such a report: it comes from the heartland of psychology. And the heartland is changing. Dr Dickinson concentrates on issues that have been central to psychological enquiry since Pavlov put them there: how do animals learn about the relationships between events that bind their environment, and themselves within it, into a causal network? How do they learn, in short, what leads to what, and what produces what? The answers Dr Dickinson gives to these questions are ones that, in some ways, Pavlov would recognize still as his own; but, in others, they would surprise him by their return to a cognitive vocabulary that he wished to expunge from the language of science. Dr Dickinson's use of this vocabulary, however, is firmly tied to experimental findings: so firmly, indeed, that (I like to think) the great Russian physiologist would soon recognize the profound identity of spirit that joins the newest developments in animal learning theory, here lucidly charted, to the experimental and theoretical tradition that he started.

For psychologists, this is essential reading. But no less so for physiologists, pharmacologists, biochemists, or anyone else struggling to understand how the brain acquires, retains, retrieves and utilizes information. Before another scientist tries to deduce something about the way memories are consolidated by stimulating, drugging or just slugging the brain after it has learned something (to take but one, if notorious, example), let him read and ponder Dr Dickinson's pages about the complex effects of post-trial surprise on animal learning. He may well then decide not to do the experiment; at least, if he does it, he will do it better; and, in any event, after his he will be profoundly sceptical of the inferences drawn by his colleagues in the past from such seemingly simple experiments.

So it is that science is most unified, not at the roots of the tree or the hub of the wheel, but at its rim; we ignore what is going on a diameter away at our peril.

TO SUSAN

'Tis sufficient to observe, that there is no relation, which produces a stronger connexion in the fancy, and makes one idea more readily recall another, than the relation of cause and effect betwixt their objects.

David Hume

Preface

The study of animal learning and conditioning in the psychological laboratory can be approached from two different perspectives. On the one hand, the psychologist can focus directly on behaviour in the attempt to formulate behavioural laws and principles which hopefully transcend the confines of the laboratory. This is the perspective of the behaviour analyst who sees the conditioning experiment as an opportunity to study in a controlled environment basic behavioural capacities which appear to be of general significance and application. The learning theorist, on the other hand, can take a very different view of conditioning. His central concern is not with behavioural change *per se* but rather with the way in which animals acquire knowledge through experience. For the learning theorist the conditioning experiment is primarily an analytically tractable tool for studying the cognitive changes that take place during learning. The relationship of the learning theorist to the conditioning procedure is very much that of a student of human information processing to his tachistoscope or the neurophysiologist to his microelectrode.

Most of the introductory books in the field of animal learning and conditioning appear to have been written primarily from the perspective of the behaviour analyst, and recent developments in animal learning theory remain largely undocumented at the introductory level. The bias has had a significant impact for it is my impression that psychologists in general assume that the bulk of work on animal learning is set within the behaviourist's framework. This volume is an attempt to redress the balance by surveying recent developments in the study of animal learning from a more cognitively orientated viewpoint. As such, I hope it will provide an introduction to the topic for both students of psychology and specialists in related areas.

The book is not intended to be scholarly in nature, and I have attempted neither to document all the relevant studies and ideas on any particular topic nor to acknowledge historical precedent. The studies to which I refer have been chosen primarily in terms of their compatibility with the general thesis of the book. This thesis, however, in no way represents a consensus view of the area of animal learning, and I apologize to those who feel I have distorted the impact of their research and ideas for my own purposes.

Finally I must acknowledge that the view of simple associative learning presented in this book has been largely shaped by discussions with G. Hall, R. A. Boakes, N. J. Mackintosh, E. M. Macphail, R. F. Westbrook and J. M. Pearce, usually by their taking issue with my arguments, and I thank them for the time they have spent trying to get my ideas straight and for commenting on an initial draft of the text. In addition, I should like to thank E. Carl for careful preparation of the manuscript and R. S. Hammans for reproducing the figures.

The data displayed in the figures have been either taken from published tables or estimated graphically from published figures.

A. D.

Cambridge
1979

1 Introduction

Animal psychology and behaviourism

The contemporary student of psychology may find it strange that the study of learning and motivation in animals should have played a central role in the development of the subject over the last fifty years or so. The original impetus behind the psychologist's interest in animal behaviour arose from the intellectual climate which fostered the birth of experimental psychology, as we know it, at the turn of the century. There were two main forces at work: the first arose from a change in the methods for studying both the mind and behaviour, whereas the second reflected the major re-evaluation of man's biological status within the cosmos brought about by evolutionary theory.

Traditionally, the main source of information about cognitive processes was through the window on the mental world provided by consciousness, or in other words by way of introspection, backed up by logical analysis and argument. However, the great achievements of the natural sciences in the eighteenth and nineteenth centuries obviously made the experimental approach attractive, and these empirical methods were imported into psychology during the latter half of the nineteenth century. They were first used in an attempt to cross the boundary between the physical and the mental worlds by formulating empirical laws relating physical variables, such as the intensity of a light source, to mental effects, such as the perceived brightness of the source. There was, however, a certain disquiet with this procedure. For the results of such investigations to be open to independent replication and investigation, all the variables involved must be publicly observable; although the intensity of the light source is a public fact, the mental impression is essentially private. If this is so, how are we to know whether the strength of a mental impression in one laboratory is the same as that in another?

This methodological difficulty could be overcome, it was argued, if our interest was shifted away from the relation between physical variables and mental effects to the study of how physical events affect behaviour, an essentially public phenomenon. And so behaviourism was born with Watson's plea that behaviour, rather than the mind, should be the true subject matter of psychology. From this perspective the behaviour of animals other than man was open to psychological investigation, and once the door was open it was soon realized that in many cases animal experiments have considerable advantages over their human counterparts. With our attention firmly focused on a rat, not only is it easier to guard against the insidious dangers of mentalism, but also the experimental method can be implemented more readily. This, of course, requires us to control all the important variables so that they can be manipulated in a precise and known manner. How could this be done when the human subject could walk out of the laboratory between testing sessions, learn and forget, love and hate, and eat and sleep in an uncontrolled fashion? Clearly a captive subject was preferable for the new methodology of behaviourism.

The second force driving psychology towards the study of animal behaviour was the growing realization of man's kindred relationship with other animals, arising from the dissemination of evolutionary theory during the second half of the last century. If human and animal morphology and physiology are related, why should not man's behaviour be illuminated by the study of other animals? It was clear that the respiratory physiologist, for example, could learn a lot about human respiration by using an animal as a model system, and many animal psychologists saw no compelling objection to employing a similar rationale in the study of behaviour, while retaining, at least nominally, the study of human capacities as his prime interest and focus.

These two trends, the empirical and the evolutionary, as well as providing the basic justification for the animal psychologist's work, also served to determine the nature of his experiments. It has often been noted, usually in a disparaging tone, that the psychologist typically employs sterile and simple experimental situations, such as the operant chamber or Skinner box and simple mazes, which rob the animal of much of its natural intelligence. He compounds this fault by studying only a restricted range of species in this impoverished environment, some of which, like the laboratory rat, are assumed to possess only a limited natural intelligence anyway. As a generalization this description is perfectly correct, but there are good reasons for adopting such an approach. Although one side of the evolutionary coin points to the similarities and relatedness of various species, the other emphasizes the specific adaptations of an animal to requirements of its own environmental niche. If, as animal psychologists, we wish to study behavioural capacities and competences which transcend particular adaptations, an attempt must be made to minimize the role of species-specific intelligence in the experimental task. The arbitrary nature of the psychologist's apparatus often represents an attempt to do just this. Although we might wish to argue about the success of the procedure and the validity of its underlying assumptions, the line of the argument is clear. A similar case can be made for concentrating the major experimental effort on a limited range of species. If we are primarily interested in capacities that transcend any particular species, rather than in variations and differences between animals, comparative studies are of interest as a test whether a given capacity is general or not. However, such comparative validation can be seen as a secondary task to that of devising an adequate theory for some target species, usually the laboratory rat and pigeon, chosen in the first instance for purely technical reasons.

So the desire for experimental control, coupled with the concept of evolutionary continuity and the belief that publicly-observable behaviour should be the basic datum of psychology, were the prime movers behind the psychologist's interest in animal capacities. The influence of theories of animal learning and motivation on the main body of the subject progressively grew from Thorndike's and Pavlov's early work in the initial decades of the century and culminated with the major impact of the ideas of Hull, Guthrie, Mowrer and Skinner in the forties and fifties. Although there are profound differences between the so-called neo-behaviourism of Hull, Mowrer, and Guthrie and the radical behaviourism of Skinner and his followers, these differences need not concern us here. The common feature uniting all these psychologists is the major influence their theories, all largely based upon animal experiments, had at the time on the whole field of general psychology. If the average American psychologist had been asked to identify the core discipline of his subject in the early fifties, he would have pointed to animal learning theory.

Over the last two decades, however, the status of the subject has been on a steady decline, until now many students of psychology often doubt whether the study of animal behaviour has any relevance to general psychology.

The attack from cognitive psychology

The seeds of the decline lay within the major theories themselves and the behaviourist philosophy which spawned them. The central ideas of the subject were so impoverished by the rigid behaviourist perspective that, when the psychological community at large shed this viewpoint, the theories died a natural death. According to behaviourism, the job of psychology is to specify the relationship between some physical event in the environment, the stimulus, and, in the case of learning, some acquired behavioural pattern, the response, without reference to mental processes. However, it turns out that most interesting behavioural capacities are just not susceptible to this type of explanation, and the neo-behaviourist theories could no more explain why the speed with which a rat runs down an alleyway increases with the size of the reward than how children learn to speak or do arithmetic.

The problem with the behaviourists' position can be illustrated by a simple and traditional animal conditioning procedure. If a rat is exposed to a series of trials in which a neutral stimulus, say a light, is paired with a noxious or aversive stimulus, such as an electric shock, not surprisingly its behaviour in the presence of the light changes. Initially the rat's activity will be hardly affected at all by the onset of the light. After a number of such trials, however, the animal is likely to show signs of being frightened when the light comes on. If it cannot escape from the experimental situation, and the shock is strong, the rat will probably crouch immobile in a hunched position when the light is presented. Clearly the animal has learned something by being exposed to the pairing of the light and shock, but the question is what? For the behaviourist, the animal has learned a new response to the light, namely crouching or 'freezing'; that is, he supposes that learning consists of the emergence of a new response. By contrast a cognitive psychologist, who is more inclined to view behavioural changes primarily as a manifestation of mental processes, is likely to argue that the rat has learned that the light predicts the shock and therefore is frightened by it. The natural response of rats to inescapable frightening stimuli is to 'freeze'. The crucial difference between this interpretation and that espoused by the behaviourist lies in the fact that for the cognitive psychologist learning consists of the formation of some novel mental structure which is only indirectly manifest in behaviour, whereas for the behaviourist the development of the new response pattern itself is learning.

It is important to realize that this distinction is not just a matter of terminology and definition. As the only evidence we have for the development of a novel mental structure in animals is through observing a behavioural change, it might be thought the cognitive view amounts to much the same thing as the behaviourist's position, at least for animals. That it does not can be illustrated by a well-known phenomenon, sensory preconditioning. Suppose that prior to pairing the light and the shock, we had exposed our rat to a series of trials in which a second neutral stimulus, a tone, was paired with the light in the absence of the shock. As both the tone and the light are stimuli with little significance for the animal, we should be unlikely to observe any major behavioural changes as a result of these pairings. Since the behaviourist defines learning in terms of a change in behaviour, he would be forced to conclude that little or no learning occurred when the light and tone were paired. Consequently, if we then go on in a second stage to pair the light with the shock, we should not expect to see any change in the behaviour elicited by the tone. In fact Rizley and Rescorla (1972), along with a number of other experimenters, have convincingly demonstrated that after the tone has been paired with the light, and the light then paired with a shock, animals are frightened by the presentation of the tone, even though the tone itself has never been presented along with the shock. For the cognitive psychologist, sensory preconditioning, at least in principle, provides no difficulty; obviously, during the first stage, exposure to the tone–light pairings set up some internal representation of this relationship which remained behaviourally silent because neither of these stimuli were of any significance to the rat at that time. When the light subsequently acquired significance by being paired with the shock, the internal structure representing the tone as a predictor of the light resulted in the tone also becoming fear inducing.

As we shall see, sensory preconditioning is but one of many examples of behaviourally-silent learning, all of which provide difficulty for any view that equates learning with a change in behaviour. Something must alter during such learning, and I shall argue that this change is best characterized as a modification of some internal cognitive structure. Whether or not we shall be able at some time to identify the neurophysiological substrate of these cognitive structures is an open question. It is clear, however, that we cannot do so at present.

I have belaboured this discussion somewhat because I think it is important to realize that being limited to observing behavioural changes does not commit you to a behaviourist perspective. There is no reason why mental processes should not be inferred from behaviour, and in fact throughout this book we shall be primarily concerned with the processes controlling the formation of cognitive structures or representations and the nature of these representations themselves. Behavioural changes will only be of interest as indices of these internal processes.

The attack from biological psychology

As I have pointed out, the learning theorist's decision to study a limited range of species intensively within unnatural and artificial situations was founded on the belief that certain learning capacities are general, at least among the higher vertebrates such as mammals and birds, and transcend the specific adaptations of any particular species. It is perfectly possible, however, that the notion of a general learning mechanism is a complete myth, and recently certain ethologists and biologically-orientated psychologists (e.g. Rozin & Kalat, 1972) have argued that the learning capacities of a given species should be adapted specifically to the particular constraints imposed by the environment. It is well known that the selective pressures imposed by the ecological niche of a species can shape both morphological and behavioural characteristics, and there is no reason why similar pressures should not have resulted in the evolution of a variety of species-specific learning mechanisms, each with its own area of application and its own principles of operation. In fact, it has been argued that the learning capacities of any particular species may be no more than an aggregate of such specific processes, and that the idea of a general mechanism which operates in a variety of species and situations is but a chimera.

There are two reasons for rejecting this conclusion. First, what evidence we have points to the existence of a general learning process. Let us illustrate this contention by looking at the recent history of taste-aversion learning in rats (Revusky, 1977). Rats are omnivorous animals and, although conservative by nature, they are prepared to eat a range of different foods and even sample new ones. In their natural surroundings this may well lead them to ingest a toxic substance. It would be to the advantage of this animal to develop a learning mechanism which is efficient in controlling aversions to foodstuffs that make it ill. Evidence for such a system was provided by Garcia and his colleagues (Garcia, Kimmeldorf & Koelling, 1955). When sickness is artificially induced in a rat, either by giving a dose of radiation or by injecting a poison, after a substance with a novel taste has been eaten or drunk, the animal will subsequently avoid food with this taste. There is nothing surprising about this result as it stands; it is well known that animals will learn about the relationship between two paired events. What initially surprised learning theorists is the efficiency of this learning. Such aversions can be learned after only a single pairing of the taste and illness even though an interval of a number of hours may elapse between the animal tasting the food and becoming ill. Traditional learning theory suggested that the animal should only learn if it became ill within a matter of seconds after tasting the food.

The first reaction to these findings was that they demonstrated a learning process which lay outside the scope of those typically studied in the laboratory, and it was noted that this mechanism had been revealed by posing the animal a more biologically orientated task. As we shall see, however, subsequent research has shown that almost every property of conventional laboratory conditioning can also be demonstrated in taste-aversion learning (Revusky, 1977), and in fact this procedure has now become one of the animal psychologist's main tools for studying learning. Rather than overthrowing the notion of a general learning process, taste-aversion research has served to change and enrich our conception of these general mechanisms.

The second reason for believing in a general learning process arises from the fact that many different animals face a common learning problem. When different species face a common environmental constraint, they often show similar adaptations in face of this environmental pressure. This similarity can occur either by the maintenance of a particular adaptation, inherited from a common ancestor, or by convergent evolution. One example of a common constraint faced by all animals that live in the sea is the relative uniformity of the medium in which they move, and faced with this universal constraint we find that many different species have similar body shapes for efficient swimming. What is the constraint that might have shaped or maintained similar learning mechanisms in a variety of species? It is of obvious importance for an animal to be able to predict when events of significance are going to occur. Survival depends upon knowing which foods make you ill and which are nutritious, which routes and paths lead to water holes, which sights, sounds, and scents predict injury and assault, and so on. Clearly animals have to possess some knowledge of the predictive relationship between events in their environment. There are two ways in which an animal may come by such knowledge: either this information can be genetically programmed in the animal's nervous system or it can be learned. Although the former mechanism may suffice in a limited environment, which is relatively constant across time, learning is required if the animal is to cope with complex and novel situations.

Given that animals should be capable of learning about predictive relationships, the question is whether such relationships have properties which are common to many different species and situations and therefore likely to maintain or shape common learning mechanisms. What type of events make the best predictors? The most obvious answer is events which lie on the causal chain leading to the occurrence of food, a mate, or a predator, for these causes are the most reliable predictors of a particular consequence or effect. Very often, however, an animal may not possess the sensory mechanisms to detect the actual proximate cause of a particular event. For instance the animal may not know about the metabolic and cellular mechanisms which lead to it becoming ill after ingestion of a poisonous substance, although it can often detect other signals and stimuli, such as the taste of the toxic food stuff, which are correlated with the underlying cause. Under these circumstances the best predictors are these indices of the underlying causal chain. So it appears that for an animal to act adaptively, it must be capable of detecting and storing information about the causal texture and structure of its environment.

It is not immediately obvious how this conclusion bears upon the generality of learning mechanisms. For instance we might well expect the evolution of specialized mechanisms since the types of causal chains encountered by a particular animal may be specific to its environmental niche. However, causal relationships have universal properties which transcend any particular instance; for example, effects never occur without a cause, and they never occur before this cause. These types of constraint may seem obvious and trivial to us, for they represent some of the most fundamental assumptions by which we make sense of the world and organize the relationships between events in space and time. Their very familiarity, however, should not blind us to the fact that they are fundamental properties shared by all causal relationships. Furthermore, any system capable of distinguishing causal relationships from chance or accidental associations must be sensitive to the unique properties of these relationships. Thus the universal nature of these properties might well lead to different animals developing and maintaining similar processes for learning about the causal structure of their environment. In conclusion, then, we can argue that from a functional standpoint there are good grounds for expecting similar learning mechanisms in a wide range of different species which can be successfully applied in a variety of situations. It is a ubiquitous requirement that animals should be capable of predicting the occurrence of events of importance to them, and the best predictors are the causes of these events, or at least detectable indices of these causes. Different species appear to show similar adaptations where the environmental constraint is universal; the homogeneity of the atmosphere coupled with the universal laws of gravity has produced similar flight mechanisms in different species, the homogeneity of the oceans along with the laws of fluid mechanics has produced similar body shapes in aquatic animals, and so we might expect the universal nature of the properties of causality to result in similar processes for learning about the relationships between events. A central assumption of this book is the existence of a basic associative learning mechanism which is common to a variety of species and designed to detect and store information about causal relationships in the animal's environment.

The implication of this argument, however, should not be overstated. There is no doubt that different species have unique ways of learning and that any particular animal has a variety of learning capacities. My argument has simply attempted to justify the case for a general process for one particular, albeit important, type of learning, namely associative learning, in which an animal just acquires knowledge about the relationship or association between events. The scope of this book will be limited to a discussion of such simple associative learning, and the theories we shall consider will have little to say about, for instance, the way in which male sparrows (Marler, 1970), or for that matter human infants, learn their natural language. Indeed, the reader should be forewarned: we shall ignore many of the central issues of general animal cognition, such as rule learning, concept formation and memory, as well as specialized cognitive abilities, such as navigation and language acquisition.

Why study animals?

The ideas and experiments covered in this book will be exclusively concerned with learning in animals. However, since we have abandoned the behaviourist's perspective, this focus may require some justification. There seem to me to be three main reasons why psychologists should be interested in animal learning over and above any general concern they may have with the cognitive and behavioural capacities of all animate beings.

The first relates to the quality and nature of the events that constitute a learning experience. Most causal chains of importance to animals, including ourselves, terminate with an event which induces an emotional or motivational state. For humans these states might range from the pleasure of a warm bath or the pain when burnt by a hot pan to the joy we can experience when we find our own affection and love mirrored by another or the anger and despair resulting from the break up of a marriage. Although these events might be the stuff of daily human life, we cannot as psychologists study experimentally what people learn from such experiences by burning our subjects, inducing marital distress, or acting as brokers for love affairs. Even if such Faustian power was within our grasp (which thankfully it is not), we should clearly not wish to exercise it. Of course, we can, as many psychologists do with great effect, turn our attention to the clinical setting and study case histories in which life has controlled the experiment for us. However, life does not perform analytic and adequately controlled studies which allow us to answer the major questions about learning. Faced with this dilemma, either we have to abandon the experimental study of the learning processes engaged by motivationally potent events or we have to turn to animal experiments. Of course, there are profound problems in generalizing between man and other animals but at least animal studies allow us to make a start on the problem. It is for this reason that much of current practice within the field of behaviour therapy was initially built on theoretical foundations laid by studies of animal learning. Although most of the theories to which the therapists initially turned were behaviouristic in nature, this should not obscure the fact that their primary relevance lies in their attempt to deal with learning about motivationally potent events.

Our use of animals, however, in no way absolves us from a moral or ethical responsibility for their well-being; if anything their total dependence upon us enhances it. The animals taking part in the studies discussed in this book have been selected over a number of years for their ability to thrive in a laboratory. In many of the experiments reported in this book the animals were not allowed continuous access to food and water but rather were fed and watered only once a day to ensure that they were willing to exhibit some behaviour at the time of testing. Such mild schedules of deprivation are probably of benefit to the general health of the animal and really do not differ from the strategy adopted by sensible dog owners in feeding their pets only once a day. Many studies also employed aversive events, such as electric shocks. In the context of the learning experiments discussed in this book, these aversive events are not used to study the reactions of animals to stress, but rather are chosen to be sufficiently mild so that they will not induce strong or chronic distress, while remaining potent enough for the animal to act on the basis of the knowledge it has about their occurrence.

One feature which often distinguishes animal from human learning experiments is the way in which the task is presented to the subject. In human learning studies the subject is explicitly instructed to achieve some goal, such as learning to keep a pointer on a moving target or memorizing a list of words, a prose passage, or story. Alternatively, the learning component of the task may be explicitly hidden from the subject, for instance, by presenting a task as one in categorizing a series of words and then at some later time unexpectedly testing for the recall of the words. Whatever the actual details of the experiment, the way in which the human subject uses different learning mechanisms and strategies will be influenced, not only by the particular events experienced within the experimental situation, but also by implicit and explicit expectations about the nature of the task and the experimenters' intentions that arise from the manner in which the experiment is presented. By contrast, in an animal study the subject is placed in the experimental environment without any explicit instructions, and we can study the way in which the sequence of events *per se* engage the learning processes. If we are primarily interested in how certain events themselves can bring about learning in the absence of expectations about the experimenter's intentions, again there are distinct advantages in using animals.

The final point is in many ways the most important of all. An obvious feature distinguishing the cognitive capacities of man from those of other animals is his ability to use explicit symbolic systems of representation. Natural language not only serves a communicative function, but also provides a medium in which man can represent and articulate his knowledge about the world. Furthermore, where natural language is inadequate, man has the ability to invent new symbolic systems to represent and manipulate his knowledge. The invention and discovery of new mathematical systems and notations are obvious examples. A feature common to all these systems is their explicit character; they can usually be spoken, written, and even in some cases programmed on a computer, and, as a result, we know a considerable amount about the way in which they serve their representative function.

Although there is little doubt that the ability to discover and use explicit representational systems is a uniquely human characteristic (but see Premack, 1976), it is not clear that this capacity necessarily implies a fundamental discontinuity between the cognitive processes of man and animal. If man stores and manipulates his knowledge about the world directly in the medium provided by these explicit symbolic systems, then it is true that there is little reason to expect any basic continuity between the cognitive powers of man and animal. However, a number of authors (e.g. Fodor, 1977) have argued that the very ability to learn an explicit language system, and to relate statements in this language to perceived events in the environment, must depend upon the possession of some more fundamental, internal and unlearned, language, which Fodor calls the 'language of thought'.

This book is not the place to evaluate the arguments for such an internal language. However, if we accept these arguments, the question of whether there are major discontinuities in cognitive processes is cast in a new light. Man's unique possession of explicit linguistic and representational powers is no longer a crucial indicator of such discontinuity. The question now becomes one of deciding whether there are any major differences in the way man and animals represent the environment in terms of the internal language. Unfortunately, we cannot begin to answer this question at present, for, as we shall see, we know next to nothing about the representational systems employed by most higher animals. What is clear, however, is that the study of animal cognition and learning may well illuminate the nature of a general, internal 'language of thought'.

Types of association

If our general contention that animals possess learning mechanisms designed to detect and store information about causal relationships is correct, then the starting point for our discussion must be with the types of relationships or associations to which they are sensitive.

Causal associations can be distinguished in terms of a number of features. For our purposes, however, two seem to be of importance; the first is the nature of the association or relationship between the constituent events, and the second the nature of the events themselves.

The nature of the association

There are basically two types of causal relationship that can hold between constituent events. Event 1, as the potential cause, can cause as an effect another event, Event 2, either to happen or not happen. The former we shall call an E1 → E2 association or relationship and the latter E1 → no E2 association. Of course, E1 need not be the immediate cause of the occurrence or non-occurrence of

E2, but could be an earlier link in a chain leading to the particular effect. In fact E1 need not even be an event in the direct causal chain, but just a detectable sign or index that the potential cause has happened. How are we to go about studying the way in which animals learn about such relationships? Obviously the only index of learning we can use is to observe a change in behaviour, and the simplest procedure is to present the animal with either an E1 → E2 or E1 → no E2 association and look for a behavioural change indicating that the animal has learned something about the relationship.

There is no doubt that animals can learn about both these relationships. Such learning can be illustrated by an experiment of Wasserman, Franklin and Hearst (1974) using hungry pigeons as subjects. The birds were placed in the typical operant chamber with two small discs situated in a wall of the chamber, one on the left and one on the right. Illumination of either of these discs with a white light for 10 seconds represented E1. The second event, E2, was the presentation of food in a small magazine placed midway between the two discs. One group of pigeons was exposed to a light → food relationship, or in other words an E1 → E2 association. Every so often one of the discs was illuminated for 10 seconds at irregular intervals, and immediately it went off food was presented. The left and right discs were illuminated equally often within the sequence of light presentations. The second group was exposed to a light → no food relationship. For these birds illumination of the discs had to appear to cause the food not to occur. This was done by presenting the food just as frequently as for the first group, but ensuring that the food never occurred when the light was on nor shortly after it had gone off.

The behaviour measured was whether the pigeon tended to approach or withdraw from the light when illuminated. Whenever a light came on, the time that a bird spent on the same side of the chamber as the light was measured and divided by the total time for which the light was on. Thus a score of 0.5 indicates that a bird spent half the time during which the light was on in the same side of the chamber as the light, whereas a score greater than 0.5 shows that the bird had a tendency to approach the light, and a score less than 0.5, a tendency to withdraw from it. Figure 1 shows these approach–withdrawal scores for the pigeons exposed to the light → food and the light → no food associations. In the case where the light could have acted as a cause of food, or at least as a signal of some underlying cause, the pigeons developed a tendency to approach the light when it was switched on. By contrast, when the light could have acted as a

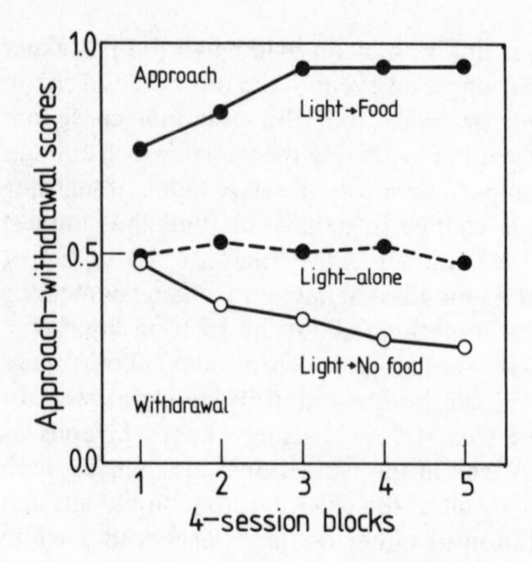

Fig. 1. The acquisition of a tendency either to approach or to withdraw from a lighted key when pigeons are exposed to different correlations between the illumination of the key and the delivery of food. An approach–withdrawal score greater than 0.5 indicates a tendency to approach the illuminated key and less than 0.5 a tendency to withdraw. In the light → food condition illumination of the key was paired with food, in the light → no food condition the food never occurred during illumination of the key nor shortly after it, and in the light-alone condition the pigeons were simply exposed to illuminations of the key but were never presented with food. (After Wasserman *et al.*, 1974.)

cause of the omission of food, the birds tended to withdraw from the light whenever it was illuminated.

These results demonstrate that pigeons can learn something about both E1→ E2 and E1 → no E2 associations when E1 is a light and E2 food. However, a number of points must be made about our interpretation of this experiment. First, we are not claiming that pigeons actually learned that the 'light caused food' or that the 'light caused the non-occurrence of food'. A discussion of what animals actually learn when exposed to these relationships must be postponed until we consider the nature of the internal representations set up by learning experiences. Second, we are not at this stage concerned with why exposure to a light → food relationship results in the animals approaching the light, and exposure to a light → no food association results in them withdrawing from it. For our present purpose all that matters is that exposure to a particular relationship produces a consistent behavioural change which we can take as an index of the fact that the animal has learned something about the relationship. Again, when we come to consider what the animal has actually learned about the association, the type of behavioural patterns we observe will become all important.

Finally, this experiment makes clear what we mean when we refer to an E1 → no E2 relationship. We can, of course, only meaningfully talk about the light as a cause of the non-occurrence of the food, or as an observable manifestation of an underlying cause, if in fact the food would have occurred had the light not been presented. This point can be illustrated by considering a third group of pigeons run by Wasserman and his colleagues. The birds in this light-alone group simply received a series of 10-second presentations of the light; food was never given in the operant chamber. In this condition the light cannot be regarded as a cause of the absence of food because the food never occurred in this situation. Causes preventing something happening can only work to counteract the effect of another cause tending to produce it. Accordingly, if these pigeons learned anything at all, it should be different from the association learned by the birds in the light → no food group. The results depicted in figure 1 confirm this analysis by showing that the birds in the light-alone condition neither approached nor withdrew from the light. An animal only has the opportunity to learn about an E1 → no E2 association in a context where it has reason to expect E2 to occur.

The problem of behavioural silence

The absence of a behavioural change does not mean that the pigeons in the light-alone condition failed to learn anything at all, and, as we shall see later, simple exposure to E1 can have profound effects on learning. The absence of any change in reactions of the pigeons to light in this light-alone condition highlights one of the major problems with taking behavioural measures as indices of learning; although we can conclude with confidence that a behavioural change shows that learning has occurred (given that we run the appropriate control conditions, about which I shall have more to say later), we cannot assume that nothing is learned in the absence of a behavioural change. In fact it will be recalled that the main reason for rejecting simple behaviourism was that learning could occur in the absence of any overt change in behaviour, or in other words that learning could be behaviourally silent.

Although learning about an E1 → E2 relationship is rarely, if ever, behaviourally silent when E2 is an event of motivational significance to the animal, such as food when it is hungry, exposure to an E1 → no E2 association often brings about learning that is not directly manifest in the animal's activity. It is easy to see why. Just as light → no food learning depends upon the light being presented in a context where the animal has reason to believe that food will occur, so the capacity of the light to elicit a behavioural response should also depend upon the presence of a similar context. Unless the animal expects food in the current situation, any knowledge it has about the light → no food association will be irrelevant. Let us consider what would happen if Wasserman and his colleagues had presented the light to the pigeons in a novel chamber which was distinctly different from the one in which they had learned about the light → no food association. As the pigeons would have had no reason to expect food in this novel setting, the light would probably have failed to produce any observable reaction, and we should then have been fooled by this behavioural quiescence into believing that the animals had learned nothing about the light → no food relationship. This means that we should always test whether an animal has learned about an E1 → no E2 relationship by presenting E1 in a context where the animal has reason to believe that E2 should occur.

Two general methods have been used for testing whether an animal has learned an E1 → no E2 association, and they can be illustrated by considering two experiments by Rescorla (1969a) on conditioned suppression in rats. In a conditioned suppression experiment a rat is exposed to an E1 → E2 association in which E1 is a neutral stimulus, such as a tone, and E2 an aversive stimulus such as shock. After a number of pairings of the tone and shock, the responses elicited by the tone change and presentation of the tone appears to put the animal in a state of fear. One property acquired by the tone is the capacity to suppress most appetitively motivated behaviour. So, if we now present the tone while the animal is pressing a lever to get food or drinking from a water spout, the rate at which the rat presses or licks decreases, and the degree of this suppression can be taken as an index of how much the animal has learned about the tone → shock relationship. Conventionally the suppressive effects of stimuli are measured by computing a suppression ratio which compares the rate of responding during the stimulus with the rate during an equivalent period immediately preceding the stimulus. It is often found that the degree to which a stimulus will suppress an appetitive response depends upon the rate at which the animal performs the response. The use of a suppression ratio attempts to take into account any individual differences in the rate of lever pressing by making the measure of suppression relative to the rate at which the animal responds prior to the stimulus. This suppression ratio normally equals $A/(A + B)$ where A is the lever-press response rate during the stimulus and B the rate prior to the stimulus. If the stimulus has no effect on responding and the rates during the stimulus and the pre-stimulus period are the same, the ratio equals 0.5; if the stimulus, however, produces suppression the ratio drops below 0.5 until it reaches a value of zero when the stimulus completely suppresses responding.

In the first phase of Rescorla's (1969a) experiments hungry rats were trained to press a lever for food in an operant chamber. After this response was firmly established, the rats were transferred to a second, conditioning chamber. Here they received a number of 2-minute tone presentations separated by 8-minute intervals. A tone → no shock relationship was established by presenting shocks at any time except during the tone itself and during a 2-minute period following each tone. A control group received exactly the same schedule of events except that a light replaced the tone. Thus this control group was exposed to a light → no shock association. In the third and final phase the rats were returned to the operant chamber and the lever-press response was re-established. The tone was then presented to all the animals for two minutes to see whether it had any effects on responding.

When he initially presented the tone to the rat which had experienced the tone → no shock association, Rescorla found that it had no effect on responding and the suppression ratio was approximately 0.5. In other words, the tone was apparently without any behavioural effect, and at first sight we might have concluded that the animals had learned nothing about the tone → no shock relationship. However, animals would not be expected to manifest in the operant chamber what they had learned about this association in the conditioning chamber. Previously they had only experienced shock in the conditioning chamber, and they had no reason to believe that shocks would occur in the operant chamber. The knowledge that the tone predicts the non-occurrence of a shock is only relevant when there are grounds for expecting shock.

The problem is how to reveal any learning which might have occurred. One way is to argue that if the rats had learned about a tone → no shock association, it should subsequently take them longer to learn about a new tone → shock relationship. So Rescorla went on to give the rats a series of tone presentations, each of which terminated with the delivery of a shock, thereby establishing a tone → shock association. These pairings were given in the operant chamber while the rats were responding for food so that he could track the rate at which the animals learned the tone → shock relationship by measuring the development of suppression across successive tone presentations. Suppression developed more slowly for the rats that had previously experienced the tone → no shock association than for those which had been exposed to a light → no shock relationship. These latter animals had never heard the tone prior to receiving the first trial in which it was paired with shock. Even though the tone failed to produce any immediate behavioural effect after the animals had been exposed to a tone → no shock relationship, they had clearly learned something about the association. This learning could be revealed in this retardation test by showing that rate at which they subsequently learned about a tone → shock relationship was retarded.

Another way of revealing behaviourally-silent learning is by a summation test. Let us suppose that after exposure to the tone→ no shock association, the animals had learned about a light → shock relationship. When we now present the light alone, it should suppress responding. However, what would we expect if both the light and tone were presented simultaneously? Under these circumstances, information about the tone → no shock association is relevant, for the animal now has reason to believe on the basis of the light that the shock will occur. Presentation of the tone should counteract the effect of the light and so reduce the amount of suppression elicited by it. Rescorla demonstrated that this is just what happens. The degree of suppression produced by the light was attenuated by presenting it in conjunction with the tone in such a summation test.

In conclusion, then, we have seen that animals can learn about both E1 → E2 and E1 → no E2 relationships. When E2 is an event of motivational importance to the animal, E1 → E2 learning is usually directly manifest in behaviour. By contrast, E1 → no E2 learning can often be behaviourally silent and fail to affect performance directly. The fact that the animal has learned something about this association, however, can be revealed by testing whether subsequent E1 → E2 learning is retarded. Alternatively in a summation test the presentation of E1 will also counteract the behavioural effects of another event predicting the occurrence of E2. We should not necessarily assume, however, that these two associations are the only relationships that an animal can learn about. Two events can be causally unrelated to each other, and animals may be able to learn that they occur independently. There is evidence for such learning, but we shall postpone any discussion of it until we have considered the conditions which are necessary for E1→ E2 and E1 → no E2 learning.

The nature of the events

In the preceding section a distinction was made between different types of association in terms of the relationship that holds between the constituent events. In this section we shall differentiate these associations in terms of the nature of the constituent events themselves. In all the examples which we have considered so far, these events have been stimulus changes in the environment, such as the onset of a tone or light and the presentation of shock or food. The cardinal feature of these is that they occur independently of whatever the animal itself does, and associations between such events reflect the operation of causal chains in the environment which do not necessarily involve the animal's own behaviour. In Rescorla's experiments, for example, the shock was paired with a light or tone, whatever the animal did. In the laboratory, learning about the relationship between such events is studied in what is called a Pavlovian or classical conditioning procedure, and all the experiments we have considered so far are examples of such a procedure. In the typical classical conditioning experiment E1, the potential cause, is usually a neutral stimulus and E2, the potential effect, is typically a motivationally significant stimulus. As we have seen, learning is measured by a change in the capacity of E1 to elicit some target response. For instance, in Pavlov's classic studies of salivary conditioning, he followed the presentation of a bell by food and measured learning by noting that the amount of salivation elicited by the bell increased.

As well as simply observing causal chains, an animal itself can act to produce changes in the environment and in this case its own behaviour is the cause of environmental events. The capacity of animals to learn about these relationships is studied by setting up an instrumental or operant conditioning procedure. This procedure programmes an association between some action of the animal, such as lever-pressing, and the occurrence of a significant environmental event, such as food or shock. When an E1 → E2 relationship is set up with the action as E1 and the shock or food as E2, we can see whether the animal learns anything about this association by observing changes in the frequency, speed, or vigour with which the animal performs the action. In the case where lever-pressing produces food, not surprisingly the frequency of this action increases if the animal is hungry, and we have an example of reward conditioning or positive reinforcement. When the lever-press produces shock, the frequency of the action decreases and a punishment procedure has been employed. We can also set up E1 → no E2 relationships with lever-pressing as E1 and food or shock as E2. When E2 is food and the action causes it not to occur, the frequency of the action declines. This is called an omission procedure. Conversely, in an avoidance or negative reinforcement procedure, the action causes the omission of a shock and its frequency increases. There is good evidence that certain animals can learn all these associations when they involve what we normally refer to as a voluntary response as E1, in that the rate at which the animal emits this action changes.

In conclusion, then, animals can learn about associations when E1 is either an environmental stimulus or a component of the animal's behavioural repertoire. My analysis, however, will be primarily concerned with learning during Pavlovian conditioning in which the potential cause, E1, is a stimulus rather than an action. This choice is determined by technical reasons that I hope will become apparent in the next chapter.

Conditioning and learning

In this book we shall study learning by using conditioning procedures, and yet the terms used to describe conditioning differ from those conventionally employed. The way in which we characterize conditioning depends very much upon what exactly we wish to study. The focus can be upon either the behavioural changes observed during conditioning or the learning processes underlying these behavioural changes.

The conventional terminology describes conditioning in terms of behavioural phenomena. In Pavlov's experiment, in which the bell came to elicit a salivary response as a result of being paired with food, the event of prime importance was this response. All the other events in the conditioning procedure are conventionally defined by reference to it. Any salivation elicited by the food is referred to as the unconditioned response because its occurrence does not depend upon the animal having been through the conditioning procedure. The food itself is referred to as the unconditioned stimulus because its capacity to elicit the salivary response does not depend upon conditioning. By contrast, the bell and the salivary response elicited by it are referred to as the conditioned stimulus and the conditioned response respectively because the capacity of the bell to elicit salivation entirely depends upon the animal having experienced the conditioning procedure. Reinforcement terminology has exactly the same emphasis. The unconditioned stimulus, the food, can also be referred to as the reinforcer because it appears to be the agent primarily responsible for strengthening or reinforcing the conditioned response.

This conditioning and reinforcement terminology would be entirely appropriate if our main interest was to characterize the behavioural changes produced by conditioning; but it is not. In this book we are concerned with changes in the animal's internal representation of its environment and the learning processes which bring about these changes, rather than with the development of new response patterns. Consequently, our description of conditioning should distinguish between the component events in terms of their potential role in the cognitive structures and in the processes controlling changes in these structures. This is what the distinction between E1 → E2 and E1 → no E2 associations attempts to do. Behavioural changes are only of interest as indices that learning has taken place and as pointers to the nature of the internal representations set up by the experience.

It might be thought that these terminological issues are of little importance, but one example can make clear the sort of confusion that can arise if we fail to maintain a distinction between behavioural and learning processes. In describing Rescorla's conditioned suppression experiment, we saw that exposing a rat to a light → shock association resulted in the light suppressing lever-pressing for food. In conditioning terminology we should describe this result by saying that the light had been established as a conditioned stimulus for suppression. In fact the light is a particular kind of conditioned stimulus, namely a conditioned excitor. A conditioned excitor is a stimulus capable of exciting or eliciting the conditioned response, in this case suppression. We also saw that, when Rescorla presented a tone → no shock association, the tone acquired the capacity to reduce or inhibit the suppression elicited by the conditioned excitor, the light, in a summation test. In this case the tone has been established as another type of conditioned stimulus, a conditioned inhibitor. The defining property of a conditioned inhibitor is its capacity to inhibit the response elicited by an excitatory stimulus.

From this example we might be led to conclude that learning about an E1 → E2 association is the same thing as establishing E1 as a conditioned excitor, and that learning about an E1 → no E2 association is equivalent to making E1 a conditioned inhibitor. However, the validity of this equation depends upon which of the target responses we focus on. In the first experiment we considered, pigeons were exposed to a light → no food association and yet the light acquired excitatory properties; it became capable of eliciting a withdrawal response. Many conditioned stimuli have both excitatory and inhibitory properties for different responses, and so identifying a stimulus as either an excitor or inhibitor is not equivalent to specifying the nature of the learning experience on which the acquisition of these properties is based.

There are two levels on which we can discuss the mechanisms involved in producing acquired behavioural changes. If our interest simply focuses on increments and decrements in the probability or strength of a particular response during a learning experience, then the conditioning and reinforcement terminology is appropriate, for this terminology makes direct reference to such changes. However, if we are primarily interested in the way in which an animal's cognitive representations of the environment change during the same learning experience, then a terminology that is likely to make direct reference to the units of this representation is more appropriate. Of course, the change in an animal's representational structure during learning acts in concert with the current motivational state of animal to produce the behavioural changes. In fact, the reinforcement mechanisms involved in establishing conditioned responses are best thought of as the interaction between the underlying processes of learning and motivation.

Three questions about learning

This introductory chapter has attempted to set the background against which we can discuss the specific processes involved in associative learning, and we can now consider what questions we wish to ask about these processes. In many ways the structure and functioning of a library provides a good analogy for the processes involved in learning. So far I have argued that learning must consist of setting up some form of internal representation of the relationship which exists between events in the animal's environment. In other words, learning is the acquisition of knowledge. Libraries are also repositories of knowledge and a simple analysis of their structure and functioning might provide us with the right sort of questions to ask about learning.

Conditions of learning

A library is formed by the acquisition of units of knowledge, called books. However, most libraries do not acquire all available books. They are selective. In any particular library most of the books might be best sellers, or originate from particular publishers, or specialize in certain areas of knowledge. We could then make a start in analysing a particular library by finding out which books are acquired and which ignored, or in other words by determining the conditions under which an acquisition occurs. Animal learning is also selective, and the first step in understanding the learning process is to specify the conditions for the acquisition of knowledge. Earlier in this chapter, I suggested that the learning processes are set up to detect and store information about causal relationships between events. If this is so, we should expect the conditions for learning to be those in which there is likely to be a causal association between the constituent events. Whether such conditions are necessary for learning is discussed in chapter 2.

Associative representations

The next question we shall address concerns the way in which knowledge is actually represented in the animal's mind. In many ways the use of a library analogy commits us to a particular perspective on this question. Many of the units of knowledge within a library, books, can be regarded as a collection of statements or facts about either the real world or, in the case of fiction, an imaginary universe. This type of knowledge is usually referred to as declarative, and is often contrasted with an alternative form of knowledge, called procedural knowledge. This distinction corresponds to that between 'knowing that' and 'knowing how'. The closest analogy to a procedural representation in our library is probably an instruction manual for performing some particular task or skill.

The question of what kind of knowledge should be represented in a declarative form and what in a procedural form is a controversial issue in artificial intelligence and human cognitive psychology (Anderson, 1976; Winograd, 1975). The procedural form tends to impose certain constraints on the way in which the knowledge can be employed. To illustrate this point, let us consider how we might express certain items of knowledge about changing gear in a car. The novice driver might receive instructions about 'changing up' in either declarative or procedural form. For instance, if he was told that 'pulling the gear lever back engages a higher gear', the information would be stated in a declarative form. The great advantage of this form is that it allows for the integration of disparate, but relevant, items of knowledge. If the novice also knew that 'the car goes faster in a higher gear', he should be able to put this statement together with the previous one to infer that 'pulling the gear lever back enables the car to go faster'. Such simple integration and inference is not possible with certain limited forms of procedural representation. For example, if the initial information has been presented in the form of an instruction 'when the engine makes a lot of noise, pull the gear lever back', there is no easy way in which the driver could put this item together with other information to derive a relationship between the speed of the car and the action of moving the gear lever. This is not to say that there are no procedural forms that permit such integration, but rather that the declarative representation is more directly compatible with integrative processes. When we discuss the problem of the way in which animals represent acquired knowledge in chapter 3, we shall approach the question of the form of the representation they employ by looking at their ability to integrate information about separate, but related, associations.

There are problems, however, with assuming that animals store information in a declarative form. These arise from the fact that we can never observe such representations directly in their passive state, but are restricted to inferring the nature of these representations from the use they are put to in controlling behaviour. It is as though we had to determine the contents of the books in our library simply by observing the actions of the readers after a visit to the library. This requires us to make certain assumptions about the processes that translate these representations into action, and questions about the form of a declarative representation can only be answered in conjunction with a plausible theory about how such knowledge is manifest in behaviour. Unless we provide such an account, we shall be open to Guthrie's famous jibe that our theories leave 'the animal buried in thought'. The problem of the interface between knowledge and action does not arise as acutely with procedural representations, for then knowledge can be encoded as an instruction which specifies the behavioural output directly. This gain, however, is bought at the expense of making it difficult, although not necessarily impossible, to account for the ability to integrate knowledge about different relationships.

The question of how animals represent simple associative information will be discussed in chapter 3.

Mechanisms of learning

Having noted that the intake of a library is selective, the next question we ask concerns the mechanisms that produce this selectivity. Perhaps the librarian just reads the best-seller list every month or so, and then purchases from the list. Alternatively, the selective intake might be due to the effective sales campaigns mounted by certain publishers. A specialized library could arise if the librarian only purchased books on the recommendation of certain individuals who share a common interest. What all these accounts are attempting to do is to describe the mechanisms controlling the intake of books. In an analogous fashion, we can investigate the nature of the mechanisms controlling the intake of information during learning. We take up this topic in the final chapter.

These three topics, the conditions of learning, associative representations, and the mechanisms of learning, do not provide an exhaustive analysis of the acquisition and storage of knowledge. Taking up our library analogy again, we can easily see that there are a number of other important questions. For instance, the position of a book within the library is usually determined by a classification system so that a borrower may easily retrieve the book he wants through the use of an index. A new acqusition must then be placed in the appropriate position within the library if it is to be of any subsequent use. Similarly knowledge acquired through learning must be located appropriately within the animal's total representational system if it is at some later date to be retrieved to control behaviour. The scope of this book, however, does not allow us to discuss in detail the structure of animal memory and the associated retrieval processes. We shall be concerned solely with the acquisition of new knowledge.

2 Conditions of Learning

In this chapter we shall discuss the conditions under which animals learn about the relationship or association between two events. An adequate description of the basic conditions of learning is an obvious prerequisite to studying both the way in which event associations are represented and the mechanisms involved in forming these representations. In the last chapter I suggested, along with a number of other authors (e.g. Testa, 1974), that associative learning mechanisms have been shaped by evolution to enable animals to detect and store information about real causal relationships in their environment. If this is so, the conditions under which learning takes place should be those in which there is likely to be a causal relationship between the events.

We have already seen that pigeons appear to learn about the association between a light and food when these events are paired with the food occurring during the light or shortly after it. Figure 2A shows the pattern of events experienced by the pigeons exposed to the light → food association in the Wasserman *et al.* (1974) experi-

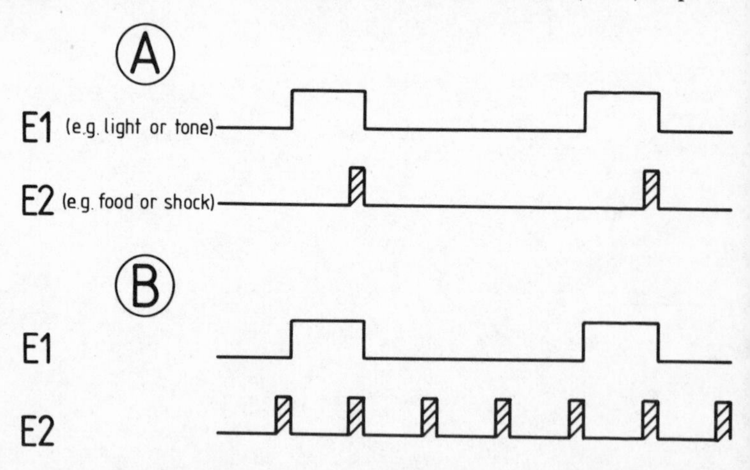

Fig. 2. The pattern of events presented when there are different correlations between E1 and E2. In Panel A E2 only occurs during or shortly after E1 and the two events are positively correlated. In Panel B E2 is just as likely to occur when E1 is absent as when it is present and the two events are uncorrelated.

ment discussed in the first chapter. The light acted as E1 and the presentation of food as E2. We saw that the strength of the approach response, and by implication the amount that the birds had learned about the relationship, progressively increased across successive pairings (see figure 1). This simple result suggests that an animal learns about the relationship whenever E1 and E2 are paired with each pairing producing an increase in the amount or strength of learning. However, if this was the only condition required for learning, the animal would be unable to learn selectively about real causal relationships in its environment. When presented with a simple pairing of E1 and E2, the animal faces the problem of deciding whether or not this pairing reflects the presence of a causal association between the events rather than a chance happening. Many totally independent events belonging to different causal chains often occur together simply by chance, and if animals learned about the association between all events which happened to be paired, they would build up a very distorted picture of the causal texture of their environment. As a result they have to use a variety of evidence in deciding whether or not a particular pairing of events reflects a real relationship. In this chapter we shall discuss the role of three sources of information which appear to have major effects on learning: the overall correlation between events, the causal relevance of the events, and their temporal relationship.

Event correlation

A central feature of the concept of causation is the idea that an effect cannot occur without an adequate cause. Thus if the E1 is the cause of E2, or an index of such a cause, E2 should not occur without E1. This is true of the pattern shown in figure 2A; at no time is E2 presented in the absence of E1. But consider the schedule shown in figure 2B. Here E2 is just as likely to occur in the absence of E1 as in its presence. Under these circumstances the pairings of E1 and E2 do not provide good evidence for a causal relationship, for the presentation of E1 fails to change the likelihood that E2 will occur, and the animal has no reason to believe that E2 would not have been presented even if the E1 had been omitted. In this case the animal should attribute pairings of E1 and E2 to chance even though such pairings happen just as frequently as in the schedule depicted in figure 2A. If the learning mechanism is to detect causal relationships successfully, the amount an animal learns about an E1 $\rightarrow$ E2 association from pairings of these events must be determined by whether or not E2 occurs in the absence of E1. Only if E2 occurs more frequently in the presence of E1 than in its absence should the animal show evidence of having learnt about an E1 → E2 association.

Rescorla (1968) performed the first systematic experiment to investigate whether the likelihood that E2 would occur in the absence of E1 affects how much animals learn as a result of being exposed to pairings of the two events. As in his experiment discussed in chapter 1, Rescorla used a conditioned suppression procedure with rats. A tone acted as E1 and a shock as E2. The animals were initially trained to press a lever for food in an operant chamber, and then were transferred to a second, conditioning, chamber without a lever and in which they received no food. Here the rats experienced a series of 2-minute tone presentations separated by intervals without the tone. For four groups the probability of receiving a shock during the tone, P (shock/tone), was 0.4, or in other words four out of every 10 tone presentations were paired with a shock. Rescorla then varied the probability that the shock would occur in the absence of the tone, P (shock/no tone). This was done by varying the likelihood that a shock would occur in the intervals between the tone presentations. For one group P (shock/no tone) was 0.4 per 2-minute interval. This means that a shock occurred in four out of every 10 2-minute intervals between the tones. In this condition there is no evidence for a causal relationship between the tone and shock, since the shock is just as likely to occur when the tone is absent as when it is present. For the remaining three groups the P(shock/no tone) was 0.2, 0.1, and 0. As the P(shock/no tone) decreases, the likelihood that there is a causal relationship between the tone and shock increases until in that last group the shock never occurs except when the tone is on. A final control group received exactly the same tone presentation as the other groups but no shocks were given. After 60 tone presentations distributed over a number of sessions, the animals were replaced in the operant chamber with the lever and the tone presented to see how much it suppressed responding.

If the extent to which the animal learned about the tone → shock association depended solely upon the number of times that these two events were paired, then all animals except the control group should have shown the same level of suppression to the tone for they all received the same number of pairings. If, however, the amount that they learned reflected how likely it was that the two events were causally related, then the degree to which the tone suppressed responding should have decreased as the probability of the shock occurring in the absence of the tone increased. The result of this experiment, shown in figure 3, obviously supported the causal hypothesis (remember that lower suppression ratios indicate more

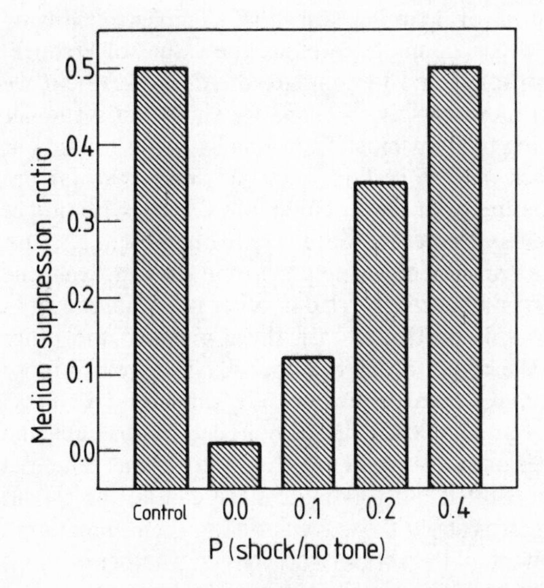

Fig. 3. The degree to which a tone suppressed lever-pressing for food after various groups of rats had received different correlations between the tone and a shock. The different correlations were brought about by holding the P(shock/tone) constant at 0.4 for all groups and varying the P(shock/no tone) across the different groups. (After Rescorla, 1968.)

suppression and more learning). When the P(shock/no tone) was 0.4, the same as the probability that the shock would occur with the tone, the animals showed no more suppression than the control group which had never experienced any shock. The amount that the animals learned about the tone → shock association increased as the likelihood of a causal relationship was increased by decreasing the P(shock/no tone). Put more formally, it appears that for an animal to learn a tone → shock relationship the P(shock/tone) must be greater than the P(shock/no tone), and that the amount it learns is related to the degree of this discrepancy. This is the same as saying that animals only learn about the relationship between two events when there is a positive correlation in time between these events, and that the amount they learn increases with the size of this correlation. When the events are uncorrelated animals do not learn an E1 → E2 association however many times the events are paired.

Overshadowing and contextual cues

At first sight these results suggest that the mechanisms underlying even this simple form of learning are very complex, involving processes which allow the animal to estimate the values of P(shock/tone) and P(shock/no tone) and to compare them. However, if we look at the situation more closely, perhaps we shall find a simpler mechanism for tracking relationships. The animal's task is to find out what caused the shock, and in certain ways our analysis so far has misrepresented the nature of this task. Up to now our presentation of Rescorla's experiment suggests that there is only one potential cause for the shock in the environment, namely the tone. In fact, from the animal's point of view, there are a myriad of other potential causes in any environment. In this particular case there are also the other stimuli provided by the conditioning chamber itself as well as those arising from the animal's own behaviour. We shall refer to these events as background or contextual cues. Looked at in this light, the animal's task becomes one of learning which of the various potential events in the environment is most likely to have caused the shock, and perhaps what it learns about the association between some target event, such as the tone, and the shock is affected by whether or not it attributes the occurrence of the shock to the contextual cues.

Even in the case where the tone and shock were perfectly correlated so that the shock only occurred in the presence of the tone, the shock was also paired with the contextual cues. If we wish to look at the way in which the presence of a second stimulus can affect learning about an association between some target stimulus and a shock, the obvious thing to do is to compare learning about this association when the second stimulus is present with that when it is absent. Unfortunately, this means that we have to pair the target stimulus with shock in the absence of any contextual cues, a feat that is, of course, impossible; there are always contextual cues present. However, perhaps we can discover the basic principles by which two stimuli interact by using as our second stimulus an event that we can control. Mackintosh (1976) recently did just this in a conditioned suppression experiment with rats using a light as the target stimulus and noises of various intensities as the second stimulus.

Mackintosh initially trained the rats to press a lever for food, and then presented a series of learning trials while the animals were pressing it. One group just received a light stimulus on each trial, whereas the remaining groups received a compound stimulus consisting of the light plus a noise. For different groups the intensity of the noise was either low (50 dB) or high (85 dB). On all trials the stimulus ended with the presentation of a shock. In order to find out how much the animals had learned about the light → shock relationship during compound training, the light was presented alone at the end of the experiment to see how much it suppressed responding. Figure 4A

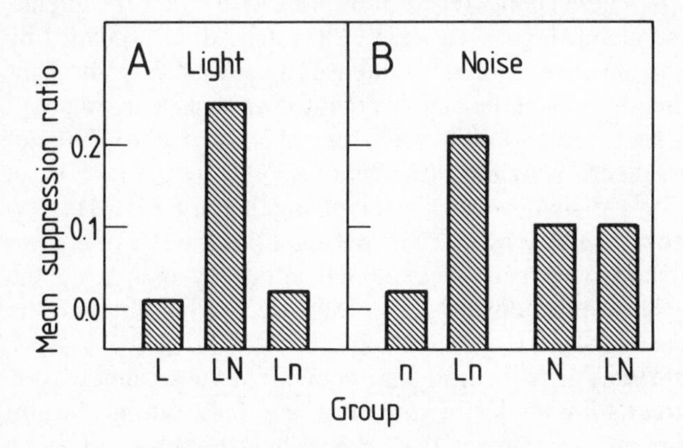

Fig. 4. The degree to which a light (Panel A) and a noise (Panel B) suppressed lever-pressing for food after different groups of rats had received pairing of a shock and the light (L), a weak noise (n), an intense noise (N), a compound of the light and weak noise (Ln), or a compound of the light and intense noise (LN). (After Mackintosh, 1976.)

illustrates the suppression ratios for the light on these light-alone test trials. The presence of an intense noise during training decreased the amount the animal learned about the light → shock association by comparison with the control group which received training with just the light. So it appears that even when the target stimulus, the light in this case, was perfectly correlated with the shock, the presence of a second stimulus decreased the amount that animals learned about the relationship between the target stimulus and the shock. This phenomenon is called overshadowing, and the noise is said to have overshadowed the light.

Figure 4A also shows that the degree of overshadowing depends upon the salience of the overshadowing stimulus. Only when the noise was intense did its presence detract from the amount the rats learned about the light → shock association. The rats showed just as much suppression to the light on test when it was trained in compound with a weak noise as when it was paired alone with shock. We might also ask whether the degree of overshadowing depends upon the salience of the target or overshadowed stimulus. In this case we need to compare the effect of the light on learning about the noise → shock relationship when the noise was intense and with that when it was weak. To assess overshadowing by the light, Mackintosh also measured the suppression produced by the noise after compound training and compared it to that maintained by the noise in two further groups. These simply received training with either the intense or weak noise alone. Figure 4B shows the suppression produced by the strong and weak noise after training. The presence of the light during training decreased the amount that the animals learned about the noise → shock relationship when the noise was weak but not when it was salient. Thus the presence of another potential E1 can detract from or overshadow learning involving the target E1. However, the degree of overshadowing depends upon the salience of the two events; overshadowing can be increased either by enhancing the salience of the overshadowing E1 or by decreasing that of the overshadowed or target E1.

There is no reason to believe that contextual cues cannot overshadow a target stimulus just as the noise and light did in Mackintosh's experiment. This means that, even when the tone and shock were perfectly correlated in Rescorla's study, it is possible that the amount the animals learned about this association was reduced from some theoretical maximum as a result of overshadowing by the contextual cues. We have no way of knowing for sure. The simple phenomenon of overshadowing, however, does not allow us to explain directly why decreasing the overall correlation between the tone and shock reduced learning. We have seen that increasing the salience of the second stimulus increases overshadowing, but the salience of the background cues remained the same whatever the correlation between the tone and shock. Perhaps, however, there are other ways of increasing the amount overshadowing, and in order to consider this possibility let us look more closely at the series of events actually experienced by Rescorla's rats.

Table 1 illustrates the types of presentations received by the rats in the correlated and uncorrelated conditions with A standing for the target stimulus, a tone in this case, and B for the background or contextual cues. All rats in Rescorla's experiment experienced a sequence of two types stimulus presentation; when the tone was presented they experienced a compound of the tone and contextual cues (A and B) and in between successive tone presentations they experienced the contextual cues alone (B alone). Animals in both conditions received the same number of pairings of the tone–context compound with E2, the shock, and the only difference was that the

Table 1. *The pairings resulting from different E1–E2 correlations*

Condition	Type of E1 presentation: A and B	B alone
Correlated	E2	no E2
Uncorrelated	E2	E2
Simple overshadowing	E2	

context alone was also paired with the shock in the uncorrelated condition. It will be recalled that in this condition the shock was just as likely to occur when the tone was absent as when it was present. Rats in the correlated condition, by contrast, received presentations of the context alone without shock. Recognition of this difference raises the possibility that these context–shock pairings in the uncorrelated condition in some way increased the ability of the contextual cues to overshadow the tone, and thus prevent learning.

Again we can investigate this idea by replacing the contextual cues by a discrete stimulus which we can directly control. A group equivalent to the correlated condition would receive trials in which a discrete stimulus B, acting in place of the contextual cues, is presented in compound with the target stimulus A and paired with E2. Intermixed with these compound trials would be trials in which B is presented alone in the absence of E2. A group equivalent to the uncorrelated condition would receive the same AB compound trials but in this case intermixed with trials on which B alone is paired with E2. We could then compare the extent to which the animals in the two conditions had learned about the A → E2 association by presenting it alone after training on one of the two schedules. If pairing B with E2 somehow increased the ability of B to overshadow A, A should elicit a weaker learned response in the group equivalent to the uncorrelated condition.

Saavedra has conducted this type of experiment (reported by Wagner, 1969a) using a rabbit eyelid conditioning preparation. In this procedure a short neutral stimulus, such as a 1-second tone or light, acts as E1 and is paired with a mild shock to the eye as E2. This shock automatically elicits an eyeblink, and when the rabbit learns about the E1 → shock association E1 also elicits an eyeblink. Saavedra ran rabbits in both the correlated and uncorrelated conditions outlined in table 1 with a light acting as A and an auditory stimulus as B. In addition a third group of rabbits received simple overshadowing training in which the AB compound was paired with shock but B was never presented alone. All animals received exactly the same number of AB–shock pairings. After this training Saavedra presented test trials on which the AB compound, A alone, or B alone was presented. Figure 5 shows the percentage of these different test

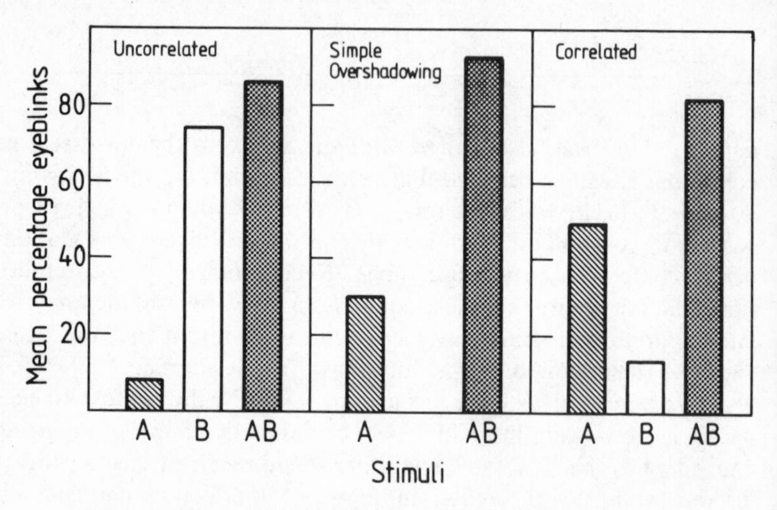

Fig. 5. The percentage of test trials on which the presentation of stimulus A, stimulus B, and a compound of stimuli A and B elicited an eyeblink after different groups of rabbits had been exposed to the uncorrelated, simple overshadowing, or correlated training schedules (see table 1). (After Wagner, 1969a.)

trials on which the stimulus elicited an eyeblink. The higher this percentage the more the animals had learnt about the association between the particular stimulus and the shock during training. The interesting result concerns the level of responding elicited by A, the stimulus equivalent to the tone in Rescorla's experiment. The rabbits in the correlated condition learned more about the A → shock association than those in the simple overshadowing group which in turn learned more than the animals in the uncorrelated condition.

Thus it appears that the B → shock pairings enhanced the ability of B to overshadow A, whereas the presentation of B alone in the correlated group attenuated this overshadowing. Why might this be so? The pattern of learning to B, which was the inverse of that shown to A, gives us a clue. The uncorrelated group appeared to learn about the B → shock association in that B elicited strong responding on test. Perhaps this learning prevented them also from learning about the relationship between A and the shock on the trials in which the AB compound was present. Within the terms of our causal perspective, we could say that if animals attribute the occurrence of the shock to B then they are less likely to attribute it also to A when A is paired with the shock in the presence of B.

The pattern of overshadowing observed by Saavedra provides a ready explanation of why Rescorla found little or no learning about the tone–shock relationship in his uncorrelated condition in spite of the numerous pairings of these two events. If we equate A with the tone and B with the contextual cues, rats in the uncorrelated condition should have learned about the context → shock association which would have allowed the context to overshadow the tone when it was paired with the shock. This argument is based upon the assumption that when the correlation between a target stimulus and shock is low the animals actually learn a context → shock association, for it is this learning which enhances overshadowing. Is there any evidence that animals actually do learn such a context → shock association under these circumstances? Dweck and Wagner (1970) studied this question using a procedure in which they measured the suppression of licking produced by a tone and by the contextual stimuli. Hungry rats were trained initially to lick a spout for a sucrose solution before receiving a number of sessions in which a tone and shock were presented in the same chamber in the absence of the spout. During each session all rats received a number of 2-minute tone presentations, half of which ended in the delivery of a shock. Thus P(shock/tone) was 0.5 for all animals. The correlated group received no further shocks, so that for these animals the P(shock/no tone) was zero and the tone and shock were positively correlated. These animals, of course, should readily learn the tone → shock association. For another group of rats, the uncorrelated group, half of each of the 2-minute periods in between the tone presentations also ended with a shock. This means that the P(shock/tone) and the P(shock/no tone) both equalled 0.5 for these rats, and the tone and the shock were uncorrelated. Following Rescorla's (1968) findings, we should have expected these animals to learn little about the tone → shock association. To find out whether this was so, Dweck and Wagner replaced the spout and re-established the licking response before presenting the tone to see how much it suppressed responding. The tone suppressed responding more for the correlated group than for the uncorrelated group, confirming Rescorla's result.

Recall, however, that the question we are considering is why do the animals in the uncorrelated condition fail to learn about the tone → shock association even though these two events are paired. The suggestion was that the presentation of shocks in the absence of the tone resulted in the animals learning about the context → shock association. This association would then enhance the degree to which the context overshadowed the tone when the compound of the contextual cues and the tone was paired with the shock. If this was so, the animals in the uncorrelated group should have learned a context → shock association, and the contextual cues themselves should have been capable of suppressing licking. Dweck and Wagner measured the degree of suppression produced by the context by seeing how long their animals took to emit the first lick when the spout was returned after the learning experience. The uncorrelated group were more suppressed, taking significantly longer to make the first lick. So it appears that in the correlated condition the animals primarily learn a tone → shock association and little about the relationship between the contextual cues and shock, whereas exactly the opposite is true of animals exposed to the uncorrelated schedule.

We started out by considering why animals learn about a relationship when the temporal correlation between the constituent events suggest that there is an underlying causal association. We can now see that this problem reduces to the question of why overshadowing occurs, and why it is modulated by the amount the animal learns about the relationship between the overshadowing stimulus and E2.

Action → E2 learning

So far our discussion has been limited to cases in which E1 is some form of stimulus change in the animal's environment, or in other words to examples of classical or Pavlovian conditioning. In the first chapter we noted that an animal can also learn about associations in which E1 is an action or response from its own behavioural repertoire. Such learning occurs within an instrumental or operant conditioning paradigm and is indexed, not by a change in the behaviour elicited by E1, but by a change in the rate or vigour with which E1, as an action, is performed by the animal. An obvious question is whether the conditions of learning we have outlined for stimulus →E2 learning, when the animal passively observes event relationships in its environment, also apply when E2 is an action and the animal interacts actively with its environment.

In fact it turns out to be very difficult to answer this question. We have seen that animals appear to learn about a stimulus →E2 association only when there is an overall positive correlation between these events. Saavedra's experiment, outlined in table 1, demonstrated that the ability to track event correlations depends upon variations in the process of overshadowing, and as a result her procedure seems to provide a way of comparing stimulus →E2 and action → E2 learning by changing the target event, A, from an external stimulus to an action generated by the animal itself. If such an experiment revealed the same pattern of overshadowing, we should have good grounds for assuming that the same conditions govern both types of learning. But as soon as we start to think about how to implement this experiment we see the problem. If the process of overshadowing operates between stimulus B and action A, we should expect the animals in the uncorrelated condition to learn little about the action → E2 association. But there is the rub. If the animals fail to learn the action →E2 association, they should not perform the target action A, and if they fail to act how can we present trials in which a compound of the stimulus B and the action A is paired with E2, as is required by the design of the experiment (see table 1), let alone equate the number of such pairings in the correlated and uncorrelated conditions. Clearly we have to equate these pairings if we wish to attribute differences in learning about the action → E2 association simply to differences in the overshadowing of the action A by the stimulus B. So it appears that the very result we expect vitiates the experiment. True instrumental conditioning procedures, involving as they often do voluntary behaviour, are not fully under the experimenter's control, and as a result it is very difficult to determine the critical conditions for learning.

There have been a number of attempts to circumvent this difficulty, all of them unsatisfactory in one respect or another. Mackintosh and I (Mackintosh & Dickinson, 1979), for instance, attempted to rob the action of its voluntary character by using a passive training procedure with rats. The design of the experiment was the same as that outlined in table 1 with A being the target action and B a tone. The action consisted of running in a wheel which could be driven by a motor so that we could force the animal to run. All the rats received a series of compound trials during which the tone was presented and the wheel driven for 15 seconds. At the end of each compound trial food was delivered to the hungry rats. During the interval between these compound trials the wheel was locked so that the animal could not run. In addition, the correlated and uncorrelated groups received a series of trials, intermixed with the tone–run compound trials, during which a tone was presented for 15 seconds while the wheel remained locked. These tone-alone trials terminated with the delivery of food for the uncorrelated group but not for the correlated group. The simple-overshadowing group received only the tone–run compound trials. After extended training on these schedules, test trials were presented without the tone in which the wheel was unlocked and not driven by the motor. These test trials measured the extent to which the animals were prepared to run freely in the absence of the tone. The more they had learned about the run → food association, the faster they should run.

If pairing the tone-alone with food enhanced its ability to overshadow running in a manner comparable to that seen with stimulus → E2 learning, we should expect the rats in the uncorrelated condition to have run slower than those in the correlated group. Figure 6 shows that this is just what happened; pairing the tone with

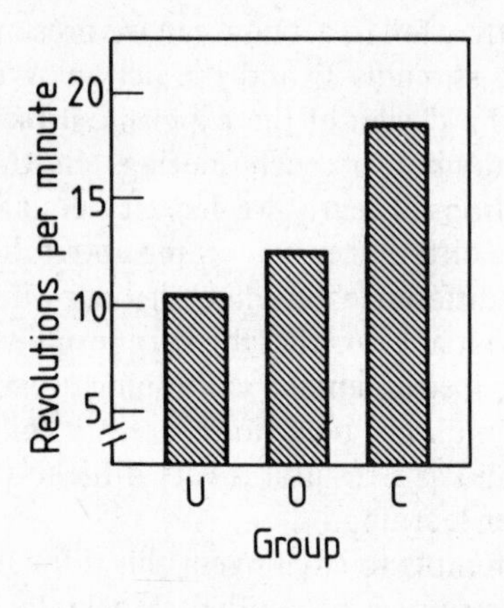

Fig. 6. The rate at which different groups of rats ran in a wheel after having been exposed to the uncorrelated (U), simple overshadowing (O), or correlated training schedules (C) with running as the target event A (see table 1). (After Mackintosh & Dickinson, 1979.)

food enhanced its ability to overshadow running in the uncorrelated group relative to the simple-overshadowing group, whereas presenting the tone in the absence of food in the correlated group decreased overshadowing. In fact the pattern of results corresponds exactly to that reported by Saavedra (Wagner, 1969a) for stimulus → E2 learning and encourages us to believe that the conditions of learning are similar in the two cases. Even so, we must recognize that the use of a passive training technique may have influenced our results. The problem of control associated with studying learning involving truly voluntary actions has meant that instrumental procedures have been largely eschewed in favour of Pavlovian or classical conditioning in which the experimenter can determine when and how frequently the different events occur. Throughout the rest of the book, I shall assume that the conditions of learning are basically the same whether E1 is a stimulus or action.

E1 → no E2 learning

We saw earlier that animals can learn about an E1 → no E2 relationship when exposed to a procedure in which E2 occurs when E1 is absent but does not occur when E1 is present. Such a relationship is likely to be one in which E1 prevents E2 or, at least, is a detectable index of an underlying cause of the non-occurrence of E2. In one of the examples we looked at, this learning was manifested in the tendency of a pigeon to move away from a stimulus predicting the absence of food. However, I also pointed out that such learning could be behaviourally silent in the sense that the stimulus might evoke no observable response when presented in certain situations. When rats were exposed to both a shock and a tone in one situation in such a manner that the tone and shock were never paired, presenting the tone in another situation did not elicit an observable reaction. The fact that the animal had learned something about the tone → no shock association could be revealed either when we tried to teach the animal a tone → shock relationship in a retardation test or when we compounded the tone with some other stimulus which predicted the occurrence of the shock in a summation test.

Just as the existence of a positive correlation between the tone and shock is a necessary condition for tone → shock learning, so the presence of a negative correlation between the two events is required for tone → no shock learning. Presenting the shock when the tone is absent means that the P(shock/no tone) has a positive value, while ensuring that the shock never occurs during the tone means that the P(shock/tone) is zero. Thus tone → no shock learning occurs when the P(shock/no tone) is greater than the P(shock/tone), or in other words when there is a negative correlation between the events.

We have already seen that the way in which a positive correlation acts to bring about tone → shock learning could be understood in terms of a more molecular analysis by taking into account the role of contextual cues. Is the same true for tone → no shock learning? If we analyse the types of events involved in a negative correlation, we find that there are basically two: first of all the contextual cues alone are paired with shock, and second a compound of the contextual cues plus the tone occur in the absence of shock. We have already seen that the way to find out how contextual cues might be involved in regulating the learning process is to replace them with an explicit stimulus which we can easily control.

We can take as our example of an experiment that does just this yet another study by Rescorla (1973) using conditioned suppression in rats. On some trials the animals received a 20-second light which always terminated with shock. In terms of our analogy the light takes on the role of the contextual cues. Intermixed with these trials were others in which a tone was presented for 40 seconds in the absence of any shock. During the last 10 seconds of each tone presentation the light was switched on. So the animals received a series of trials in which pairings of the light and shock were intermixed with trials in which the light–tone compound was presented in the absence of shock. If this training procedure is sufficient to allow the animals to learn a tone → no shock association, the animal should come to discriminate between trials in which the light is presented alone and those in which the light is presented in compound with the tone. When the light is presented alone, the animals should suppress responding having learned a light → shock association; however, if they have also acquired a tone → no shock association, presentation of the tone in compound with the light should reduce the degree of suppression. This is, of course, a variety of the summation test discussed earlier. Figure 7B shows that on test trials after training, the compound of the tone and light did in fact maintain less suppression than the light alone. So the basic condition for learning a stimulus → no shock relationship is one in which the stimulus is presented in the absence of shock, but along with another event which has been paired with the shock. The obvious implication is that a negative correlation between a stimulus and shock induces stimulus → no shock learning because it presents these particular pairings with the contextual cues playing the role taken by the light in Rescorla's experiment.

There is, however, something puzzling about this result. Why should the animals have learned a tone → no shock association when they are in fact exposed to pairings of the tone and light. Instead they might have learned about the relationship between the tone and light. When discussing the limitation of the behaviourist's approach in the first chapter, it was pointed out that animals are perfectly capable of

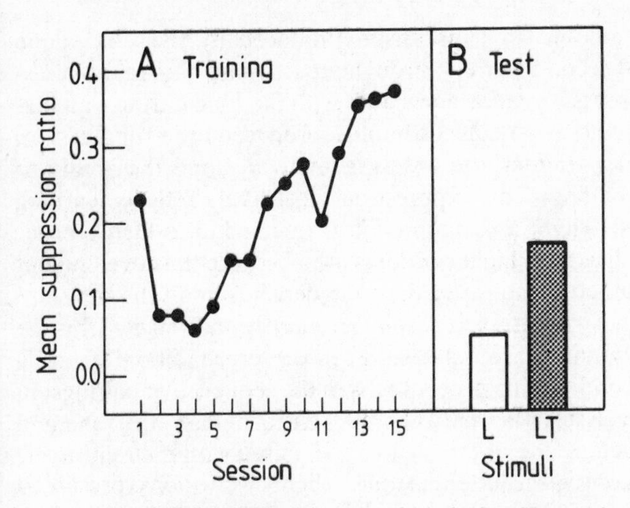

Fig. 7. Panel A: The degree to which the first 30 seconds of 40-second tone presentations suppressed lever-pressing for food by rats when the last 10 seconds of the tone was compounded with a light which was independently paired with shock.
Panel B: A comparison of the extent to which the light and a compound of the tone and light suppressed responding after the training illustrated in Panel A. (After Rescorla, 1973.)

learning about the relationship between two neutral stimuli, such as a tone and light, from simply being exposed to pairings of these two events. This phenomenon is called sensory preconditioning. It will be recalled that the tone → light association was revealed by pairing one of these stimuli with a shock whereupon presentation of the other then induced suppression. A similar effect might be expected in the present experiment. The tone and light were paired, and so the animals should have learned a tone → light association. Since the light, as a predictor of shock, became a fear-inducing stimulus, so should the tone as a predictor of the light. There is only one major difference between the sensory preconditioning procedure and that used in the present experiment; in the former all the tone–light pairings occurred before one of the elements was associated with shock, whereas in the latter case these events are intermixed. However, there is no obvious reason why this should matter. Consequently, we might expect the tone to maintain suppression through its association with the light, rather than alleviating suppression induced by the light. In fact, initially the tone did just this. It will be recalled that on tone–light trials the onset of the tone occurred 30 seconds before the onset of the light, and so Rescorla was able to measure the amount of suppression produced by the tone alone during this 30-second period throughout training. Figure 7A illustrates the suppression ratios associated with the tone during training. Initially, the tone, as a neutral stimulus, produced little suppression, but with further training the tone gradually acquired the ability to suppress responding. This suppression most likely reflects learning about the tone →light association and is referred to as higher-order conditioning. The term higher-order is used because the development of the conditioned response to the tone depends upon the development of another conditioned response, namely that elicited by the light. The only difference between a sensory preconditioning and a higher-order conditioning procedure is in the sequence of pairings; in the former the association between the neutral stimuli, the tone and light, occurs before one of the elements is paired with the reinforcer, the shock, whereas in the latter the tone→ light association is presented after (or at the same time as) that between one of the elements and the shock.

What is really interesting, however, about the results displayed in figure 7 is that the tone gradually lost its capacity to suppress responding with extended training. This probably reflects the fact that learning about the tone →light association gradually became dominated by learning about a second relationship, the tone → no shock association, so that by the end of training the tone, rather than producing suppression itself, was actually capable of reducing the suppression maintained by another stimulus. Thus we see that exposure to even these relatively simple sequences of events in controlled laboratory conditions can lead to complex patterns of learning, and we can begin to appreciate the difficulties in trying to find out exactly what animals have learned in natural environments where the number and temporal sequences of events are much richer.

A molar view of the conditions for learning

Let us try to summarize at this point what we have learned so far in a more abstract and general form. Basically animals appear to learn an E1 → E2 relationship when the P(E2/E1) is greater than the P(E2/no E1), and an E1 → no E2 relationship when the P(E2/E1) is less than P(E2/no E1). Although the way in which animals actually make contact with the overall correlations is through the actual pairings of events, including the contextual cues, brought about by the correlations, there are some advantages in representing the conditions for learning at a more molar level.

Figure 8 illustrates a space in which lie the points defining every possible combination of P(E2/E1) and P(E2/no E1). If there is any

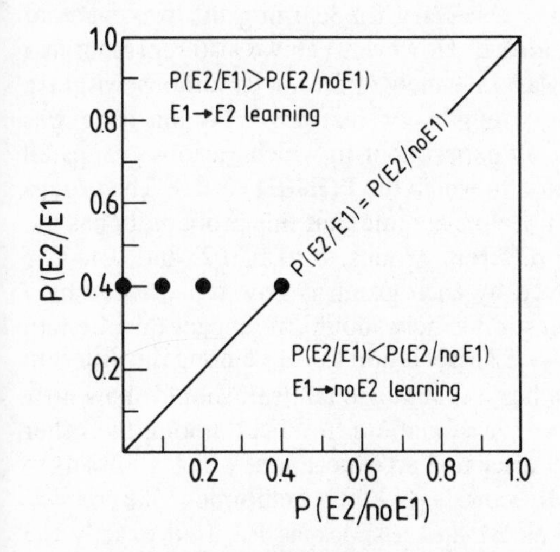

Fig. 8. A contingency square representing possible combinations of the conditional probabilities of E2 occurring when E1 is present, P(E2/E1), or absent, P(E2/no E1). Along the diagonal E1 and E2 are uncorrelated and associative learning does not appear to occur. Above the diagonal E1 and E2 are positively correlated and E1 → E2 learning occurs, whereas below the diagonal the two events are negatively correlated and E1 → no E2 learning is observed. The filled circles represent the conditions run in Rescorla's (1968) experiment (see figure 3).

simple relationship between the position of a point defining an event relationship in this space and the type and strength of learning which results, this representation would obviously provide a succinct way of summarizing the conditions necessary for learning. Along the diagonal the P(E2/E1) equals the P(E2/no E1) and E1 and E2 are said to be uncorrelated or to occur randomly with respect to each other. There is now considerable evidence that exposure to conditions represented by points along this diagonal does not lead to E1 acquiring the capacity to elicit a behavioural pattern indicative of either E1 → E2 or E1 → no E2 learning. This is true however often E1 and E2 are paired or E2 occurs in the absence of E1. Above the diagonal line, the P(E2/E1) is greater than the P(E2/no E1) and the events are positively correlated so that the animals should learn something about the E1 → E2 relationship. By contrast, below the diagonal line where the P(E2/E1) is less than the P(E2/no E1) the events are negatively correlated and the animals should learn an E1 → no E2 association.

So we can see that the two sectors above and below the diagonal line specify the conditions necessary for learning the two types of association we have considered. However, can we also represent in a systematic fashion the way in which different conditions within a sector lead to different strengths of learning? To illustrate this problem, let us consider an experiment in which various groups all receive a learning condition in which the P(E2/E1) is 0.4. The groups differ in the P(E2/no E1). Let us assume that this probability has the following values for the different groups: 0, 0.1, 0.2, and 0.4. The learning condition received by each group is now represented by a point in figure 8. The question is how should we expect the strength of learning about the E1 → E2 association to vary among the different groups. The group which lies on the diagonal itself should show little learning. But are there any systematic differences among the other groups? We have already discussed an experiment which allows us to answer this question. Rescorla's (1968) conditioned suppression experiment with a tone as E1 and a shock as E2 used exactly the probabilities illustrated in figure 8. The results of this experiment are illustrated in figure 3 and show that as the P(E2/no E1) decreased, the amount of suppression, and by inference, the strength of learning increased. The problem now is whether there is any simple feature of the position of a condition in figure 8 that determines how much the animal will learn. Unfortunately, the simple answer is that we do not as yet know, for no experiments have systematically investigated the whole of this space. However, there is an attractively simple hypothesis; the further away a point is from the diagonal line in the direction of the upper left hand corner, the stronger is learning. As they stand, Rescorla's results fit this general hypothesis, for as the P(E2/no E1) decreases the distance of a condition from the diagonal line increases. At present, we can only predict the relative strength of learning in different conditions according to this hypothesis, when all the conditions have either the same P(E2/E1) or the same P(E2/no E1). We have no idea about the shape of the lines that connect conditions of equal learning throughout this space (for further discussion of this problem see Gibbon, Berryman and Thompson, 1974).

At this point in our discussion, a couple of important asides should be made. The first concerns the concept of 'strength of learning', which so far we have used fairly freely. As we have often pointed out, the only way that we can know that animal has learned, for example, a tone → shock relationship is by observing a change in the tone's behavioural properties, usually its capacity to suppress responding. So, if as a result of being paired with the shock the tone suppresses responding, we infer that the animal has learned about the tone → shock relationship. However, we have also gone one step further; we have assumed that the more the tone suppresses responding, the stronger is learning about the tone → shock association. This may seem reasonable. We find that increasing the number of pairings between the tone and shock increases the suppressive capacity of the tone, and attribute this to stronger learning about the tone → shock association. But we have also seen that we can manipulate the degree to which the tone will suppress responding, not only by changing the number of tone–shock pairings, but also by varying the likelihood that the shock will occur without the tone. By saying that both these manipulations affect the strength of learning, we are claiming that they both act on the learning mechanism in a similar manner. To put this another way, the claim is that when the tone elicits the same degree of suppression in two groups of animals they have learned the same thing about the association, even though they might have been exposed to a different number of pairings between tone and shock and a different overall correlation between these events. Intuitively this may seem a very unlikely assumption. Why should the animals exposed to the different correlations not have learned different things about the relationship between the tone and shock? It might just happen that with the particular number of pairings received by each group the same behavioural effects occur. Of course, this is a possibility. However, our assumption that variations in the number of pairings and changes in correlation have a common effect on something we call the 'strength of learning' is more plausible when we remember that the way in which changes in overall correlation actually effect learning is via overshadowing by the contextual cues. Decreasing the overall positive correlation between a tone and shock decreases the effectiveness of tone–shock pairings in bringing about learning because it potentiates overshadowing by the contextual cues. So, when we actually look at the molecular conditions of learning, it is perfectly reasonable that changes in correlation and the number of pairings have the same effect; both represent changes in the number of potent or effective pairings.

The second aside also concerns the problem of inferring learning from behavioural changes. For instance, when we pair a tone and shock, we assume that the change in the behavioural properties of the tone reflects learning about the tone → shock relationship. But how do we know that such changes are really due to this association? Such pairings, as well as presenting the animal with the relationship between these events, have two other effects; they ensure that the animal is simply exposed to both the tone and the shock. It is well known that simple exposure to a stimulus can change an animal's responsiveness to it, a phenomenon called habituation. In addition, exposure to strong stimuli, like shocks, can alter the animals responsiveness to other stimuli. These latter effects are usually referred to as sensitization and pseudo-conditioning. If these changes occur without an association between the tone and shock, is it not possible that the changes we see when we present a relationship are due simply to exposure to the events rather than to learning about the relationship? In order to assess whether a behavioural change reflects true associative learning, we should really run a control condition in which a group of animals receives just as many presentations of the tone and shock but in the absence of an association. Rescorla (1967) has suggested that such a control condition should lie at the appropriate point on the diagonal of figure 8 so that the two events occur just as frequently as in the experimental group, but randomly with respect to each other. He referred to this condition as the truly random control condition. If we now find that the tone elicits a stronger response in a group which receives the tone and shock paired by comparison to the appropriate truly random control group, we can be confident that this effect is due to learning about the tone → shock association. The only difference between the paired and random groups is the relationship between the tone and shock; both groups receive an equal number of exposures to the stimuli, so that any sensitization or pseudo-conditioning should occur equally in the two groups.

Blocking and surprise

Animals learn about an E1 → E2 association when there is a positive correlation between the events. This finding accords with the idea that animals learn whenever E1 and E2 are paired so long as we recognize that such learning can be attenuated when E1 is overshadowed by other stimuli. The degree of overshadowing depends upon whether or not these stimuli are independently paired with E2. Why might this be so? Figure 9 shows a possible sequence of four trials experienced by animals in the correlated and uncorrelated conditions from table 1 with A as the target E1 and B as the overshadowing E1. We know from Saavedra's results (see p. 33) that extended exposure

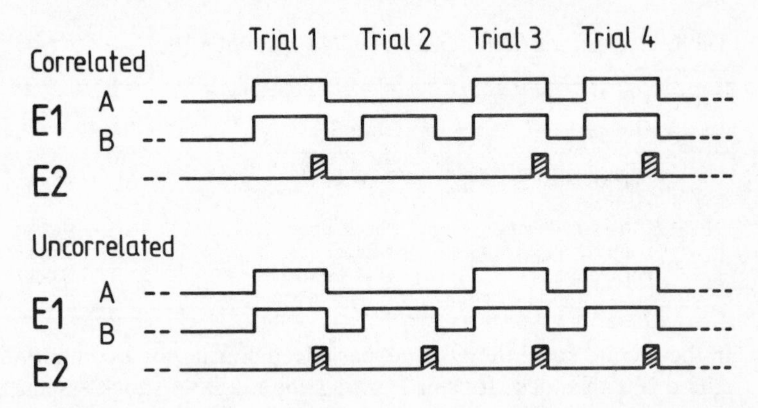

Fig. 9. An illustration of a possible sequence of four trials experienced by animals exposed to the correlated and uncorrelated schedules of table 1. On Trials 1, 3, and 4 all animals experienced pairings of a compound of A and B with E2. On Trial 2, however, B alone is paired with E2 in the uncorrelated schedule, while E2 is not presented in the correlated condition.

to this type of sequence will lead to the correlated group learning more about the A → E2 association than those in the uncorrelated condition. If we assume that learning about this relationship only occurs when A and E2 are paired, the difference between the two groups can occur only on Trials 1, 3 and 4. It cannot occur on Trial 1, however, for the two groups have received exactly the same experience by the end of Trial 1, and therefore the difference must arise from variations in the amount learnt on Trials 3 and 4. By Trial 3 the only factor separating the two groups is that the animals in the uncorrelated condition have received a prior pairing of B and E2 on Trial 2 and thus had an opportunity to learn about the B → E2 association. Perhaps if the animals have already learned something about the B → E2 relationship, the ability of B to overshadow A on subsequent compound trials is enhanced.

There is now abundant evidence that this is so. We can illustrate the effect by yet another experiment by Rescorla (1971) using a conditioned suppression procedure with rats in which A was a light, B a tone, and E2 a mild shock. The design of this two-stage experiment is illustrated in table 2. The designation of the various groups in this table refer to the relationship between A and the shock across the two stages of the experiment. After the rats had been trained to press a lever for food in an operant chamber, they were transferred to a conditioning chamber for Stages 1 and 2. In the first stage the uncorrelated group received a number of pairings of B and a mild shock in order that they might learn about a B → shock association.

Table 2. *Design of Rescorla's (1971) experiment*

Groups	Stage 1	Stage 2
uncorrelated	B → shock	AB → shock
simple overshadowing	shock-alone	AB → shock
control	B/shock	AB → shock
correlated	B → no shock	AB → shock

In the second stage they then experienced pairings of a compound of A and B with shock. If prior learning about a B → shock association enhanced the extent to which B overshadowed A during compound training, these animals should have learned little about the A → shock relationship in the second stage. To see whether this was so, Rescorla ran two further groups for which B was not paired with the shock during Stage 1, although they received the same number of compound trials in Stage 2. The simple-overshadowing group experienced the same number of shocks as the uncorrelated group during Stage 1 but were not presented with B, whereas the control group received the same number of both B and shock presentations distributed randomly throughout each session. As we have already pointed out, Rescorla has argued that such a truly random control condition represents the appropriate group against which to assess any effects of experiencing an association between B and the shock.

Following Stage 2, Rescorla re-established lever-pressing in the operant chamber and then presented A to see by how much it suppressed responding. As figure 10 shows, the uncorrelated group showed less suppression than the overshadowing and control groups indicating that prior learning about the B → shock association had enhanced the extent to which B overshadowed A. Kamin (1969), who was the first to demonstrate this effect, referred to it as 'blocking' because pre-training to B in Stage 1 appeared to block learning about A in Stage 2. This blocking effect lies at the heart of the ability of animals to track event correlations and it is obviously important to determine why it happens. After considering a number of possibilities, Kamin (1969) eventually came down in favour of the idea that animals only learn an E1 → E2 association if a surprising or unexpected event occurs at about the same time as E2 is presented. In a simple learning situation where a stimulus is paired with a shock, the occurrence of the shock itself is surprising or unexpected, at least during the initial trials, in the sense that neither the stimulus nor the

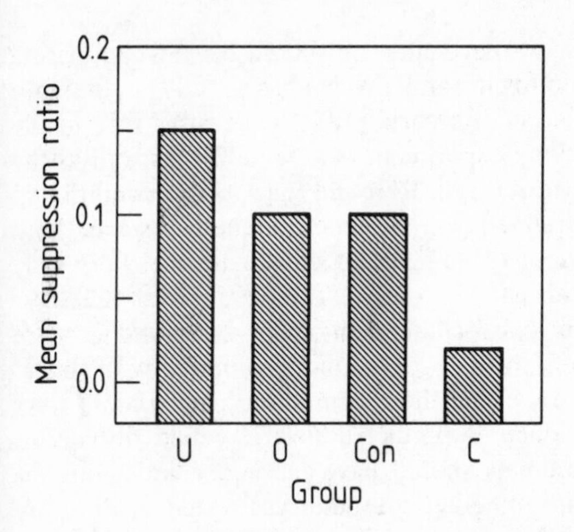

Fig. 10. The degree to which stimulus A suppressed lever-pressing for food after the rats had been exposed to pairings of an AB compound with shock. Prior to compound training the different groups of rats had experienced various associations between B and shock (see table 2). The uncorrelated group (U) received prior pairings of B and shock, the simple overshadowing group (O) was just pre-exposed to the shock alone; the control group (Con) experienced random presentations of B and shock, and correlated group (C) experienced a negative correlation between B and the shock. (After Rescorla, 1971.)

contextual cues predict its occurrence. It is the surprising nature of the shock that brings about learning. By contrast, when the AB compound is paired with the shock in Stage 2 of a blocking experiment (see table 2), the occurrence of the shock is entirely predicted by the B in the uncorrelated condition and hence not surprising. The shock is predicted by B because in Stage 1 the animal has learned a B → shock association, and as the occurrence of the shock in Stage 2 is fully predicted, the animal fails to learn about the A → shock relationship. It is a small step from this idea to the suggestion that the amount an animal learns during a series of pairing of two events depends upon how surprising or unpredicted the second event, E2, is.

This surprise hypothesis also has implications for the way in which we could go about enhancing learning. If somehow we could make the occurrence of E2 super-surprising, we should get super-learning. The occurrence of an event should be most surprising, not when it is presented in the absence of a predictive stimulus, but when it occurs in the presence of another stimulus which actually predicts it will not happen. This means that a particular E2 should be most surprising during or immediately following an E1 which has already entered into an E1 → no E2 association. Rescorla (1971) tested this idea in the second half of his blocking experiment (see table 2). As well as the groups we have already discussed, Rescorla ran rats in a condition in which A was highly correlated with the shock. In the first stage, these rats received presentations of both B and shock with the shock only occurring when B was absent. As a result, these animals should have learned a B → no shock association. In Stage 2, A and B were presented in compound, and the compound was paired with shock. The occurrence of the shock on these compound trials should have been very surprising, since B predicted that it would not occur. Consequently, these animals should have learned a lot about the A → shock relationship. Whether they did was tested in the third stage by presenting A alone to see how much it suppressed lever-pressing for food. Figure 10 shows that the light suppressed responding more in this correlated group than in the simple overshadowing and control groups for which B had been established neither as a predictor of the shock nor as a predictor of its absence. Rescorla referred to this enhancement of learning as superconditioning. The whole pattern of Rescorla's results supports the idea that the amount an animal learns about the relationship between a neutral stimulus and a shock depends upon how surprising the shock is. When it is fully expected, blocking occurs, and when it is very surprising, we observe superconditioning.

We have been through a labyrinth of argument in attempting to analyse why animals learn about event relationships only when there is a correlation between the events. In summary, the ability to track the overall correlation between events can be analysed in terms of variations in the capacity of contextual cues to overshadow the target E1. These variations in overshadowing are brought about by processes akin to those seen in the blocking procedure. The problem of why blocking occurs in turn reduces to the question of why E2 has to be surprising if the animal is to learn an E1 → E2 association over a series of pairings. This question is the central concern of the discussion of mechanisms of learning in the final chapter.

Learned irrelevance

In the last section I concluded that when two events occur randomly with respect to each other, or in other words are uncorrelated, animals learn neither an E1 → E2 nor an E1→ no E2 association, and furthermore it was suggested, along with Rescorla, that such a random condition was the appropriate control against which to assess whether learning about either of these two associations had in fact occurred. This conclusion does not imply, however, that animals learn nothing when exposed to a random relationship, and it well might be that they learn something which reflects the fact that the events are causally unrelated. We have already seen that any such learning must be behaviourally silent, for the responses elicited by a tone are not obviously changed when it is presented in a random relationship to shock. The problem is to decide how to make any such learning behaviourally active.

Mackintosh (1973) has argued that if an animal learns that two events are unrelated, it should be harder for it to learn about any subsequent association between them. Baker and Mackintosh (1977) have recently investigated this possibility with thirsty rats, using a tone as E1 and the delivery of water from a drinking spout as E2. In the first stage the uncorrelated or random group were given a series of sessions in which the tone and water were presented in a random relationship. Three other control groups received an equal number of presentations of the tone alone, or the water alone, or neither of these events in the first stage. In the second stage all the groups were exposed to a tone → water association. On each trial the tone came on and 10 seconds later water was delivered. The amount the animals had learned about the tone → water association was measured by recording the time for which the animals made contact with the drinking spout during the 10 seconds of the tone prior to the onset of the water. This time was compared to that for which contact was made during a 10-second pretrial period prior to the onset of the tone. If the animals had learned about the tone → water association, they should have contacted the spout more when the tone came on. This increase was measured simply by subtracting the pretrial contact time from the tone contact time. Figure 11A illustrates these difference scores for the four groups during the second stage when the tone was paired with the water. All four groups showed an increase in the relative contact time during the tone, indicating that they had learned the tone → water association. The interesting result, however, was that the rate of this learning was retarded in the random group. If animals are exposed to a schedule in which the tone and water occur independently, the rate at which they subsequently learn a tone → water association is retarded.

The problem is how to interpret this retardation. We have already seen that if animals have the opportunity to learn an E1 → no E2

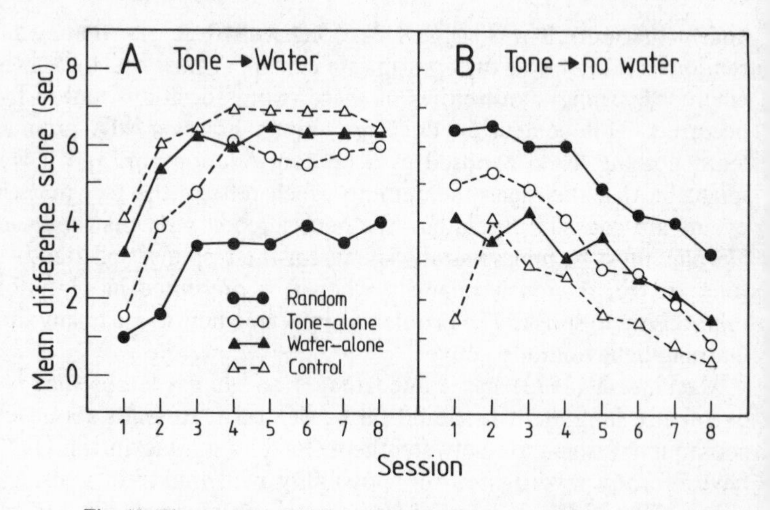

Fig. 11. The extent to which various groups of rats contacted a licking spout during a tone stimulus when the tone was either paired with the delivery of water to the spout (Panel A) or presented in compound with a light which was independently paired with water (Panel B). The difference score represents the difference between the time for which the animals contacted the spout during the tone on a trial and the contact time during an equivalent period immediately preceding the tone. Prior to this training the random group experienced uncorrelated presentations of the tone and water, the tone-alone group simple pre-exposure to the tone, and the water-alone group pre-exposure to water, while the control group experienced neither event. (After Baker & Mackintosh, 1977.)

association, they will then be slower to learn about an E1 →E2 relationship. The obvious implication is that animals, given independent presentations of tone and water, learn something similar to a tone →no water association. This means that if the animals were exposed to a tone → no water relationship in the second stage, the random group should learn more rapidly about this than the control groups. Baker and Mackintosh investigated this possibility in a second experiment. The first stage was exactly the same as in the previous study; one group was presented with random tone and water presentations, one with water alone, one with tone alone, while the final group received neither water nor tone presentations. In the second stage all the rats were exposed to a tone → no water association. On some trials a second stimulus, a light, was paired with water, and these trials were intermixed with others in which a tone–light compound was presented in the absence of any water. The rate at which the animals stopped contacting the spout during the tone–light compound trials should provide a measure of their readiness to learn about the tone → no water association. If animals in the random condition learn something like a tone → no water association in the first stage, the rate at which they learn when this relationship is explicitly presented in the second stage should be enhanced.

In fact exactly the opposite result was found. Figure 11B shows the difference score on the tone–light compound trials during the second stage of this study; the random group actually contacted the drinking spout for a relatively longer time during these trials, suggesting that the rate at which they learned about the tone → no water association was also retarded. Taken together these two studies show that exposure to random presentations of the tone and water hinders subsequent learning about both tone → water and tone → no water associations. One interpretation of this finding, favoured by Mackintosh (1973), is that animals can learn that two events are independent, or in other words that they are causally unrelated. Subsequently this learning will retard the acquisition of information about any causal relationship which might exist between the events. Unfortunately we cannot be certain of this explanation for there are plausible alternatives. For instance, figure 11 shows that simply presenting either the water alone or the tone alone in the first stage will retard acquisition in the second, although not as much as in the random condition. Whatever the reason for these pre-exposure effects, and we shall have more to say about them later, the greater retardation shown by animals exposed to both the water and the tone may just be a summation of the simple effects produced by presentation of each event alone. If this is so, clearly the retardation in the random group cannot be taken as conclusive evidence that the animals learned that the two events occur independently. So far no one has thought of a way of disentangling this problem.

The experiments by Baker and Mackintosh and other similar studies have been directed at finding out whether animals can learn that two specific events are unrelated. Certain authors (e.g. Maier & Seligman, 1976) have claimed that animals can show a much more general type of irrelevance learning. Work in this area has concentrated on instrumental conditioning procedures in the attempt to demonstrate that animals can learn that some environmental event occurs independently of anything they do. The normal way of demonstrating such learning is also by looking for a retardation effect when the animals are subsequently exposed to an E1 → E2 or an E1 → no E2 relationship, just as Baker and Mackintosh did. However, instead of employing an environmental stimulus as E1, these studies look at learning when E1 is an action. There are now a large number of experiments of this type, and I shall describe only a couple of studies to illustrate the basic effects.

Jackson, Maier and Rapaport (1978) initially trained two groups of rats to press a lever for food in an operant chamber. Then the rats were placed in a small restraining tube during the first stage. Here the random group received a series of shocks which occurred independently of anything the animal did. This procedure was designed to give the animals an opportunity to learn that the shock could not be controlled by their activity, and that its occurrence was causally unrelated to any aspect of their behaviour. A control group was simply restrained in the tube for the same period. In the second stage the rats were once again placed in the operant chamber and lever-pressing was rewarded with food. In addition the animals were punished for lever-pressing with a shock whenever a noise stimulus was present, or in other words a lever-press → shock relationship was in operation whenever the noise was on, but not when it was off. The more the animals learned about the lever-press → shock association, the more the presentation of the noise should have been capable of suppressing responding relative to a period when the noise was absent. The degree of this suppression was expressed by the normal suppression ratio, and figure 12A illustrates these ratios for the two groups. Clearly the rats in the random group which had been pre-exposed to the shock in the tube, showed less suppression than the control group. This difference is not due to the fact that shock pre-exposure simply reduces its aversiveness. A second set of rats was given exactly the same pre-exposure in the tube in the first stage, but instead of having to learn a lever-press → shock association they were exposed to a simple noise → shock association in a conditioned suppression procedure. Whereas in the punishment study the delivery of the shock during the noise depended upon the animal responding, in the conditioned suppression procedure the shocks were paired with the noise whatever the animal did. Figure 12B shows that there was no difference between the level of suppression maintained by the noise for the random and control groups when the animals were required to learn simple noise → shock association. A difference would have been expected if pre-exposure had affected punishment simply by reducing the aversiveness of the shock or by increasing the rats' tolerance.

Jackson, Maier, and Rapaport take the difference in the punishment condition as indicating that rats can learn that the shock occurs independently of anything that they do, or in other words that the

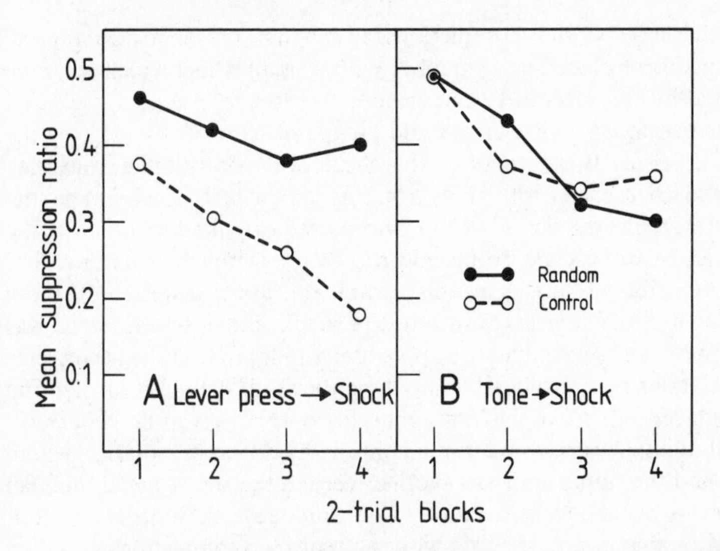

Fig. 12. The suppression of lever-pressing by rats for food produced by a noise either when lever-pressing was paired with shock only in the presence of the noise (Panel A) or when the noise was paired with shock independently of lever-pressing (Panel B). The random group had been pre-exposed to unsignalled and inescapable shock in a different apparatus, whereas the control group had received no such pre-exposure. (After Jackson *et al.*, 1978.)

occurrence of the shock is uncontrollable. Presumably, during pre-exposure the rats produced a number of responses in an attempt to avoid, minimize, or in other ways cope with the shock. Consequently, when the occurrence of the shock could actually be controlled by the rat's behaviour during the punishment procedure, the pre-exposed animals took longer to learn about the lever-press → shock relationship because such learning was at variance with the information they acquired during the first stage.

If this interpretation is correct, at least two predictions follow. First, if the animals really learn that the shock occurs whatever they do, such learning should delay the acquisition of information not only about action → shock associations but also about action → no shock relationships. This means that pre-exposure to uncontrollable shock should retard the acquisition of an avoidance or escape response, as well as reducing the effectiveness of punishment. The acquisition of an avoidance and escape response is based upon learning about an action → no shock association. Second, the retardation effect should be abolished if shocks received in the first stage were in fact under the animal's control. Another experiment by Jackson and his colleagues provides a test of these predictions. In the first, pre-exposure stage all the rats were placed in a chamber with a small wheel which could be rotated by the forepaws. One group, the escape group, was given a series of shocks which they could escape or turn off by rotating the wheel. So for these animals, the shock was controllable, and they learned to escape from it by rotating the wheel. Each rat in the second group, the yoked group, was paired or yoked with one of the animals in the escape group and received exactly the same number and duration of shocks as this escape rat. As a result, both these groups had the same exposure to the shock; the only difference was that the yoked group had the opportunity to learn that the occurrence of the shock was unrelated to anything they did. A final control group was placed in the wheel-turn apparatus but was given no shocks.

All animals were trained in a completely different situation where they had the opportunity to avoid or escape the shock by learning an action → no shock association. The apparatus, a shuttlebox, consisted of two compartments with a small interconnecting tunnel so that the animals could cross freely from one compartment to the other. Every so often a shock was presented which could be terminated by crossing from the compartment in which the animal was at the time into the other one, and then back again into the

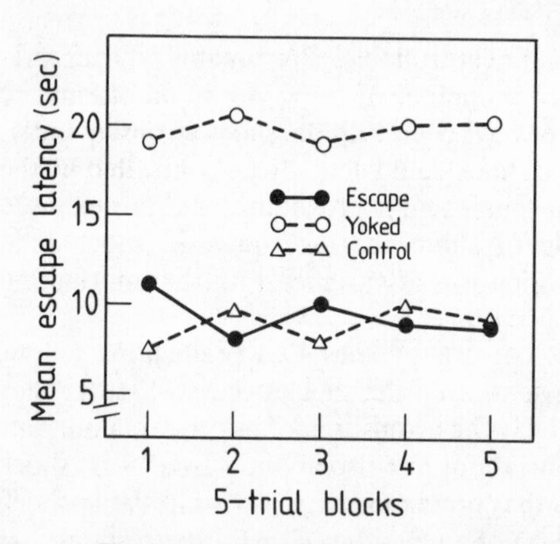

Fig. 13. The latency of escape from a shock in a shuttle box after the various groups of rats had been exposed to different conditions. Prior to training in the shuttle box the escape group had been trained to escape from shocks in a wheel-turn apparatus, whereas the yoked group had received exactly the same number and temporal pattern of shocks without being able to escape them. The control group received no pre-exposure to shock. (After Jackson *et al.*, 1978.)

original one. So an effective escape response required two crossings. Figure 13 shows the time it took the rats to make these escape responses after the onset of the shock. The animals in the yoked group, pre-exposed to uncontrollable shock, exhibited much greater escape latencies than the control group. Taken together with the previous study, this demonstrates that whatever is learned during the pre-exposure phase, it subsequently interferes with learning both action → shock and action → no shock associations. The retardation effect does not occur, however, if the shock was under the animal's control during the first, pre-exposure stage; the escape group emitted the shuttle-escape response just as rapidly as the control group. Maier and his colleagues (Maier & Seligman, 1976) have taken this type of retardation effect, which they call learned helplessness, as evidence that an animal can learn that the occurrence of a particular event is causally unrelated to anything that it does. Although alternative accounts have been offered (see Glazer and Weiss, 1976), the evidence for learned helplessness is compelling.

This concludes the discussion of the effects of event correlation on learning. We have seen that animals can learn about both E1 → E2 and E1 → no E2 associations, and that when exposed to these associations they appear to learn something different in each case. Exposure to one will retard learning about the other. In addition, animals appear to be able to learn that E1 and E2 are independent, or at least in certain cases that no E1 is related to E2. Again this type of learning is different in the sense that exposure to random occurrences of E1 and E2 will retard the acquisition of information about both E1 → E2 and E1 → no E2 associations. Event correlations are one of the defining criteria of a causal association and determine the nature of that relationship. However, the presence of a causal association can be indicated by a number of other properties, and if the learning mechanism has evolved to detect such a relationship, we should expect them to be sensitive also to these secondary properties.

Causal relevance

In the real world causal chains are indicated, not only by the overall correlation between the constituent events, but also by the nature of the events themselves. Certain types of causes are likely to produce certain types of effects. As an example, let us suppose that you suddenly find yourself feeling particularly nauseous after a visit to an exotic restaurant. If the illness only occurs after visiting this restaurant, you are likely to attribute the cause of your distress to the food you ate there. However, a hidden assumption is buried in this conclusion. The decor of the restaurant is just as well correlated with illness as is the nature of the food, and yet we are unlikely to attribute the illness to the colour of the wallpaper (except, of course, in exceptional circumstances). Why might this be? Obviously because taste stimuli, produced by the ingestion of food, are much more likely to be closely associated with the underlying causal chain producing gastric distress than are visual stimuli produced by the decorations. There are qualitative constraints operating between cause and effect.

Certain of these constraints are fairly general, and the taste–illness relationship we have considered is likely to be one. If learning mechanisms are designed to detect and store information about causal relationships, an animal should more readily learn about a taste → illness association than the relationship between say a tone or light and illness. Domjan and Wilson (1972), among others, have looked at whether rats show such selective learning.

Two groups of rats received a 35-second oral infusion of saccharin. Immediately after this infusion, one group was injected with a toxin, lithium chloride, to induce sickness, and the other with saline to act as a control. Another pair of groups was given a 35-second exposure to a buzzer followed by an injection of either lithium or saline. If the animals were more ready to learn a taste → illness association than a buzzer → illness relationship, they should have been less ready to perform a response which produced the taste than one which produced the buzzer. After three pairings of either the taste or the buzzer with the appropriate injection, all animals were given a preference test in which two spouts were available to find out whether selective learning had occurred. For all groups one spout contained water. The second contained saccharin for the groups that received the saccharin solution during training and water for the groups which had received the buzzer. In this latter case, each time the animal made contact with the second spout the buzzer sounded. Thus the more the animals had learned about the stimulus → illness association, the more they should prefer the first spout containing water alone.

Figure 14A shows the percentage of fluid taken from this first spout containing water alone. When saccharin was paired with saline injections during training, the rats took most of their fluid from the saccharin spout; this bias reflects the rat's initial preference for sweet solutions. By contrast, when the saccharin had been paired with lithium chloride, the animals clearly avoided the saccharin and consumed most of their fluid from the water-alone spout. No such

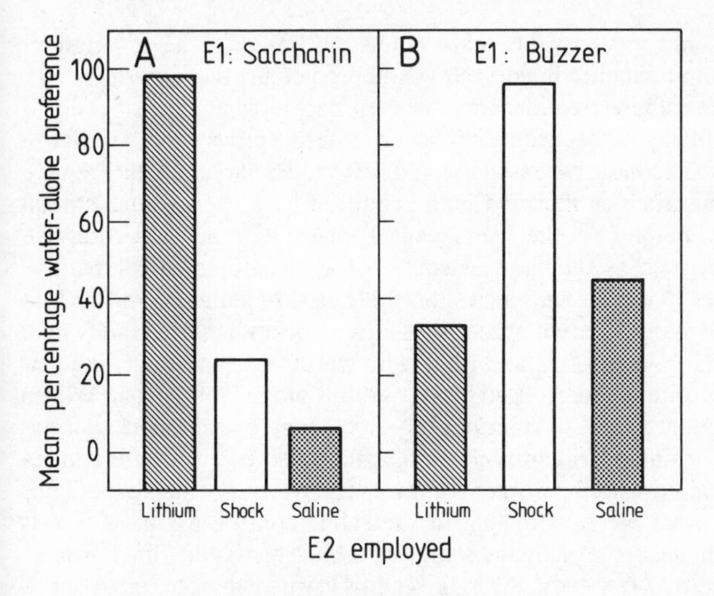

Fig. 14. The preference of rats for drinking from a water spout when the animals were given a choice between this water spout and either one delivering a saccharin solution (Panel A) or another water spout which activated a buzzer when the animal made contact (Panel B). The various groups received prior training in which either the taste of the saccharin solution (Panel A) or the buzzer (Panel B) was paired with the induction of illness by an injection of lithium chloride, a shock, or a simple saline injection. (After Domjan & Wilson, 1972.)

avoidance emerged when the stimulus paired with the toxin was the buzzer; the rats took roughly the same percentage of fluid from the water-alone spout irrespective of whether the buzzer was paired with lithium chloride or saline.

The implication of these results is clear; rats are more ready to learn about saccharin → illness association than a buzzer → illness relationship. But why is there such a difference? Perhaps the saccharin solution is a more salient stimulus for the rat, for it is well known that the rate of learning varies with the salience of E1. If this is the explanation, rats should learn saccharin → E2 relationships more rapidly than buzzer → E2 relationships whatever the nature of E2. By contrast, my causal hypothesis argues that the difference arises not from the properties of the saccharin solution and buzzer taken alone, but from the fact that the induction of illness is more likely to be associated with the same causal chain as that producing taste stimuli. If we choose as E2 rather than the saccharin, an event likely to belong to the same causal chain as stimulus such as the buzzer, the rats should more readily learn about the buzzer → E2 association. The simple salience hypothesis would predict just the opposite result.

To test these two ideas we have to decide what type of event is more likely to be produced by an exteroceptive stimulus, like a buzzer. The intuitive answer would seem to be another exteroceptive stimulus, such as the mild pain produced by a shock. Such stimuli usually belong to the same causal chain as other exteroceptive stimuli, such as the sight or sound of a predator or a competitive member of one's own species. So the causal hypothesis predicts that animals should learn a buzzer → shock association more readily than a saccharin → shock relationship, whereas the salience theory predicts the opposite outcome. Two further groups of the Domjan and Wilson experiment received either buzzer–shock or saccharin–shock pairings during training. In addition, both groups also received saline injections paired with the buzzer and saccharin to equate their experience of injection *per se* with that of the saline control groups. The rats were then given exactly the same preference test as the other groups. As figure 14B shows, the rats readily learned the buzzer → shock association, but showed little knowledge of the saccharin → shock relationship. Pairing the buzzer with the shock led to a marked increase in the percentage of fluid consumed from the water-alone spout relative to the saline controls, whereas pairing the saccharin with shock had little effect on the animals' initial preference for the saccharin spout. This is exactly the opposite preference pattern from that found when the stimuli were paired with lithium chloride.

These and related results obviously support the idea that learning occurs selectively in favour of associations which are likely to reflect an underlying causal relationship. However, it should be pointed out that this type of experiment is difficult to interpret, and great care has to be taken in designing the studies to ensure that the results are really due to selective learning. Domjan and Wilson ran certain procedural controls in order to rule out alternative explanations. For instance, this pattern of results could have arisen if the simple fact of being ill shortly before the preference test caused the animals to avoid salient tastes, irrespective of the stimulus paired with the lithium chloride. To rule out this possibility, Domjan and Wilson gave all groups preference tests with both the saccharin and the buzzer, although we have only described the relevant results. Pairing the buzzer with lithium chloride did not make the animals avoid the saccharin on test. In addition, they equated the extent to which all animals were exposed to both the buzzer and the saccharin during training; the stimulus which had not been paired with the lithium chloride or shock was presented 90 to 150-seconds after each pairing. Even so, with sufficient ingenuity it is possible to think up explanations for these results which do not appeal to selective learning (see LoLordo, 1979). However, at present, we shall take this experiment as pointing to the fact that learning occurs selectively in favour of causally related stimuli. This conclusion is strengthened by the fact that other factors, which might be expected to reflect the likelihood of a causal relationship between E1 and E2, affect learning in a systematic manner. So, for instance, E1 → E2 learning is more rapid both when E1 and E2 occur in the same rather than disparate spatial locations (Rescorla & Cunningham, 1979; Testa, 1975) and when they are in the same modality (Rescorla & Furrow, 1977).

One final point must be made about the selectivity of learning demonstrated in this type of experiment. As we mentioned in the first chapter, initially this selectivity was taken as evidence against a general learning mechanism. It was argued that such selectivity pointed to a fundamental difference between the mechanism's underlying taste-aversion learning and learning about the association between exteroceptive stimuli. However, almost all the phenomena typical of laboratory conditioning preparations have now been demonstrated with taste-aversion learning. E1 → no E2 learning, sensory preconditioning, higher-order conditioning, blocking and superconditioning have all been shown with taste-aversion procedures (Revusky, 1977). As a result, it is difficult to resist the conclusion that the processes underlying saccharin → illness and buzzer → shock learning are the same, and that the selectivity we observe should be interpreted as a general property of a common mechanism. And, of course, if this general mechanism is designed to detect causal relationships, as the work on event correlation in conditioned suppression suggests, such selectivity is just what we should expect.

Temporal relationships between events

We saw earlier that there had to be an overall positive correlation between a potential cause, E1, and the effect, E2, for an animal to learn an E1 → E2 relationship. However, the operation of this correlation could be reduced to that of the individual pairings of both E1 and E2 and of the contextual cues and E2. The phenomena of overshadowing and blocking would then explain why E1 → E2 learning only occurred in the presence of an overall positive correlation. However, there is a hidden assumption behind this analysis, namely that for learning it is necessary that the two events occur close to each other in time. This is what we mean when we talk about a pairing. In addition, in all the cases of successful learning we have considered so far, the onset of E1 has occurred before the onset of E2. The question we want to consider now is whether such forward pairings and temporal contiguity are necessary conditions for learning.

The obvious approach to this problem is to perform an experiment in which we explicitly vary the time interval between, for example, a tone and shock to see whether it has any effect on how much the animals learn about the tone → shock association. There have been many experiments of this type, and we shall just consider one by Mahoney and Ayres (1976) to illustrate the basic pattern of the results. Mahoney and Ayres gave rats a single presentation of a 4-second tone and a 4-second shock and varied the interval between the onset of these two stimuli for different groups. For three groups the onset of the tone preceded that of the shock by 0, 4, 8, or 150 seconds, whereas for a further two groups the onset of the shock preceded that of the tone by 4 or 8 seconds. A final control group just received the shock presentation. This control group was actually run in another experiment but the procedure of the two studies was almost identical. On the next day a conditioned suppression test was used in which the tone was presented while the rats were thirsty and licking a spout for water. The time taken for the rats to complete 10 licks during the tone was measured. The more the animals had learned about the tone → shock relationship, the greater would be the degree of suppression and hence the longer should be the latency to complete the 10 licks.

Figure 15 illustrates these latencies as a function of the interstimulus interval between the onset of the tone and the onset of the shock. Obviously the time interval between the two events has a profound effect on learning. The animals learned most about the tone → shock association when the onset of the tone preceded that of the shock by about 8 seconds, and learning deteriorated when the interval was either decreased or increased. In fact when the onset of the tone preceded that of the shock by 150 seconds or the shock occurred before the tone, the animals showed little evidence of learning by comparison with the control group. The poor learning observed when the tone actually comes on after the shock appears to be entirely congruent with the causal hypothesis. It is a cardinal feature of our intuitive concept of causality that the effect should not precede the cause, and so the ineffectiveness of backward pairings might be

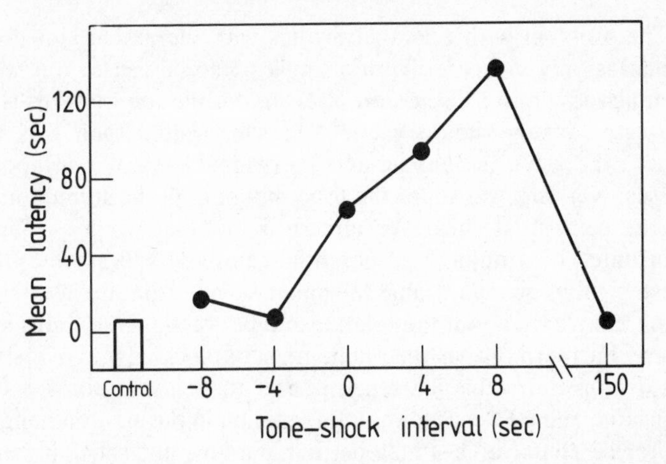

Fig. 15. The suppression of licking by rats produced by a tone after different groups of rats had experienced a single pairing of a tone and shock with various intervals between the onset of the tone and the shock. The control group just received a presentation of the shock but no tone. The degree of suppression was measured by recording the latency to complete 10 licks in the presence of the tone on test; the longer the latency the more severe the supression. (After Mahoney & Ayres, 1976.)

expected. However, we should remember that animals are actually faced with the task of learning about the relationship betwen perceived indices of cause and effect, not necessarily the events themselves, and the criterion of temporal ordering does not apply as strictly for such indices. If an animal is attacked from the rear, it may not see the aggressor until after the attack. However, clearly the animal should learn from such an experience so that it is in a position to take appropriate action if confronted by the same aggressor in future. So we can see that there might be some functional grounds, however tentative, for expecting animals to learn something about an E1 → E2 relationship even when the onset of E2 precedes that of E1. However, in terms of our causal hypothesis, other aspects of these results are in many ways surprising. There is no immediately obvious reason why learning should be so drastically affected by the time interval when the tone precedes the shock. Many causal chains can extend over long periods, and it may well be that only the initial links in such a chain can be directly perceived. The detection of long causal chains should be just as important to animals as knowing about short ones, and, assuming that the function of such learning is to provide the animal with a predictor of events of importance, it might well be an advantage to learn about the prior links in the chain.

The problem with this analysis lies with the assumption that the animal is only presented with a single potential cause. But we have seen already in our discussion of event correlation that this is never the case. When Mahoney and Ayres presented their rats with a shock, the environment actually provided a number of potential causes. Not only was there the tone, but also all the stimuli provided by the contextual cues. We already know that when an animal is confronted by a number of potential causes, it will assume that the most highly correlated and the most salient stimulus is the actual cause, and learn about the relationship between this stimulus and the effect. Furthermore, as the phenomena of blocking and overshadowing demonstrate, this learning appears to detract from the amount which the animals learn about other stimuli in the environment. Once we realize that even a simple pairing of a tone and shock involves, in addition to the tone, contextual cues as potential causes of the shock, the question of how the process of overshadowing affects learning with different tone–shock intervals arises.

The degree to which a target stimulus is overshadowed by other stimuli depends upon two factors: the relative salience of the target and overshadowing stimuli and the amount the animal learns about the association between the overshadowing stimuli and E2. Perhaps the overshadowing can be similarly modulated by varying the time interval between the target stimulus and E2. For instance, we could assume that at the optimal interval for tone → shock learning the tone is functionally equivalent to a very salient stimulus, and that it will overshadow the context. At this interval the presence of the contextual stimuli will have little effect on learning about the tone. As we deviate from the optimal value of about 8 to 10 seconds, the tone becomes functionally equivalent to a stimulus of relatively low silence. This will have two effects. First, it will enable the contextual cues to overshadow the tone, and thus decrease the amount that the animals learn about the tone → shock relationship. This effect will account for the basic interstimulus interval function displayed in figure 15. Second, it will no longer allow the tone to overshadow the context, so that the animals should show an increase in the amount they learn about the context → shock relationship.

An obvious prediction follows from this account. If we varied the time interval between the tone and shock and, instead of measuring tone → shock learning, recorded context → shock learning, we should find that little learning occurred with a tone–shock interval of about 8 to 10 seconds, but that the amount the animals learned about the context → shock association increased as the time interval departed from this optimal value. In other words, the pattern of context → shock learning should be the exact inverse of the pattern for tone → shock learning displayed in figure 15. Unfortunately Mahoney and Ayres did not measure the amount the animals learned about the context → shock association, but Odling-Smee (1975) has done so in a somewhat similar experiment. Odling-Smee trained his rats in an apparatus consisting of a small black compartment interconnected with a large white compartment. In the first stage the animals were placed in the small black compartment and given 10 trials in which a tone was paired with a shock. For different groups, the interval between the onset of the tone and the delivery of the shock was varied between 3.3 and 90 seconds. In order to measure how much the animals had learned about the relationship between the contextual cues provided by the black compartment and the shock, the rats were simply placed in the apparatus for 300 seconds with a door between the white and black compartments open. To the extent that the animals had learned about the black compartment → shock association, they should spend most of the 300 seconds in the white chamber. Figure 16 shows how much time each of the groups

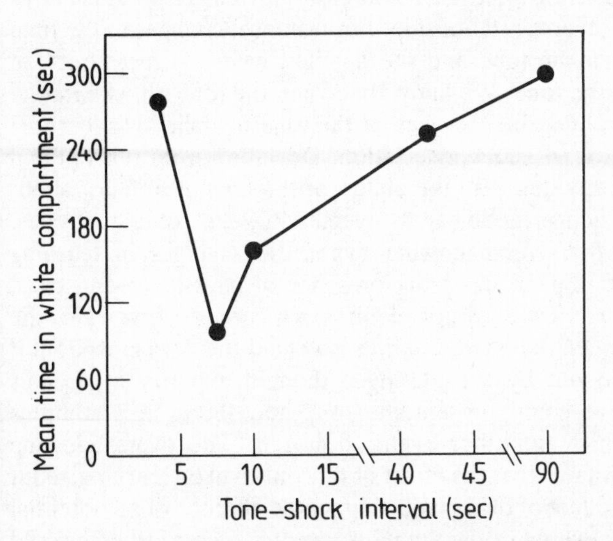

Fig. 16. The time rats chose to spend in the white compartment of a two-compartment apparatus after having experienced a series of tone–shock pairings in the other, black compartment. The interval between the onset of the tone and the shock during the pairings was systematically varied across different groups of rats. The greater the time the rats spent in the white compartment on test, the more they had learned about the association between the black compartment and the shock. (After Odling-Smee, 1975.)

receiving different tone–shock intervals spent in the white compartment. When the tone–shock interval was 10 seconds, the rats spent most of the 300 seconds in the black compartment, and appeared to have learned very little about the relationship between these contextual cues and the shock. Their preference for the black compartment reflects the rat's innate attraction to dark places. As the tone–shock interval deviated from 10 seconds in either direction, the animals spent progressively more time in the white compartment indicating increased learning about the black compartment → shock association. A comparison of figures 15 and 16 shows that there is an inverse relationship between the amount learned about tone → shock and the context → shock associations; the more the animals appear to learn about the context, the less they learn about the tone → shock association.

This inverse relationship is clearly in line with the idea that deviations from an optimal tone–shock interval of about 8 to 10 seconds decreases learning about the tone through an increase in the ability of the contextual cues to overshadow the tone. However, these results in no way provide conclusive support for the overshadowing hypothesis, and we could easily formulate an alternative account of the inverse relationship. For instance, changes in the time interval between the tone and shock might have a direct effect on learning about the tone. We know that when the tone–shock interval is about 10 seconds, the presence of the tone overshadows learning about the context → shock association. Deviation from this optimal value may directly decrease the ability of the animal to learn about the tone, and hence its ability to overshadow the contextual cues. According to the overshadowing hypothesis changes in learning about the tone depend upon the presence of overshadowing cues, while the second view does not require such cues. At first sight, the difference between these two theories may be difficult to grasp, but it can be brought out by considering a thought experiment. Let us suppose that we could present the tone and shock in a stimulus vacuum, in which no other events occurred. The overshadowing hypothesis maintains that the effect of time interval on learning about the tone is a product of overshadowing; in the absence of all potential overshadowing events in the stimulus vacuum such a process could not operate, and the strength of tone → shock learning should be unaffected by variations in the interval between the two events. By contrast, if changes in the tone–shock interval have a direct effect on learning, the absence of potential overshadowing cues should not matter and we should see the same decline in learning as the interval varied from the optimal value.

Such an experiment is, of course, impossible for we can never create a complete sensory vacuum. At the very least, there will always be events provided by the animal itself. However, perhaps we could minimize the extent to which the contextual cues are likely to overshadow the target stimulus. Contextual cues are predominantly stimuli provided by the animal's external environment, and thus primarily occur within the visual, auditory, and tactile modalities. Such stimuli are readily associable with events like shock, and thus they are always likely to exert a strong overshadowing effect when the animal attempts to learn about the relationship between some target stimulus and a shock. However, we have also seen that rats do not readily learn about the relationship between exteroceptive stimuli and the induction of illness. Consequently such contextual cues should only exert a weak overshadowing effect when an animal attempts to learn about the association between a taste and illness. Taste–illness pairings occur within a sort of functional stimulus vacuum, because the type of contextual cues provided by most environments are not causally relevant to the induction of illness.

If we are correct in supposing that variations in the time interval between two events affect learning through the overshadowing process, we should expect such variations to have far less effect in the case of taste → illness learning than in the case of tone → shock learning. There is now considerable evidence that this is the case. For instance, Smith and Roll (1967) gave thirsty rats 20 minutes access to a saccharin solution followed by exposure to X-radiation of sufficient magnitude to induce illness. For different groups the interval between the end of the period of access to saccharin and exposure to X-radiation was varied between 0 and 12 hours. The development of any aversion to saccharin was measured 48 hours later by giving the rats a choice between licking a spout delivering saccharin and one delivering water. Figure 17 shows the percentage of fluid consumed from the saccharin spout for the different groups. The preference of control groups, just receiving prior exposure to the saccharin without the X-radiation, is also plotted. The most striking feature of figure 17 is the length of the interval over which animals associated the taste and illness. Even with an interval of six hours, the animals still showed a significant aversion to the saccharin by comparison with the control group. This finding stands in marked contrast to the failure of rats to learn about the relationship between a tone and shock from a single pairing when the interval is as short as a couple of minutes. Although this discrepancy might be partially due to differences in the

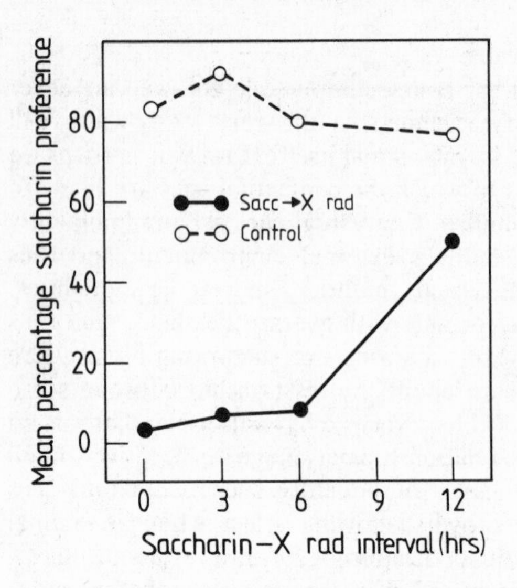

Fig. 17. The preference of rats for drinking from a spout delivering a saccharin solution rather than a water spout after different groups of rats had experienced a pairing of the intake of saccharin with the induction of illness by exposure to X-radiation. The interval between access to saccharin and the induction of illness was varied across the different groups. Rats in the control condition simply received access to saccharin without the induction of illness. (After Smith & Roll, 1967.)

relative salience of the tone and taste and of the shock and illness, it is unlikely that animals can relate events such as a tone and shock when the interval between them is a matter of hours, whatever the strength of the tone and shock.

This discrepancy, however, is just what we should expect if overshadowing plays a major role in determining the interval over which the association between two events will be learned. In the case of tone → shock learning, the contextual cues are highly associable with the shock, and as a result deviations from the optimal tone → shock interval will result in strong overshadowing of the tone. By contrast, in taste-aversion learning the contextual cues are not usually readily associable with illness and the degree of overshadowing should be weak. Revusky (1971) was the first to point out that overshadowing might play a different role in these two situations. He went on to argue that if this type of explanation is correct, we should be able to reduce the effective interval for taste-aversion learning by interposing relevant stimuli between access to saccharin and the induction of illness. In an attempt to test this idea, Revusky gave a number of groups of rats access to 2 ml of saccharin solution followed one hour later by an injection of lithium chloride to induce illness. In addition, he allowed the animals to consume 5 ml of a second solution 15 minutes after they had finished the saccharin. The salience of this second solution, and hence its potential capacity to overshadow the saccharin, was varied by giving the different groups exposure to a vinegar solution of concentrations varying between 0 and 4.5%. Finally he measured the degree of aversion to the saccharin in three daily preference tests. In each test the animal had a choice between a spout containing the saccharin solution and one containing a novel caffeine solution.

Figure 18 illustrates the percentage of fluid consumed from the saccharin spout on the second test. As the concentration of the

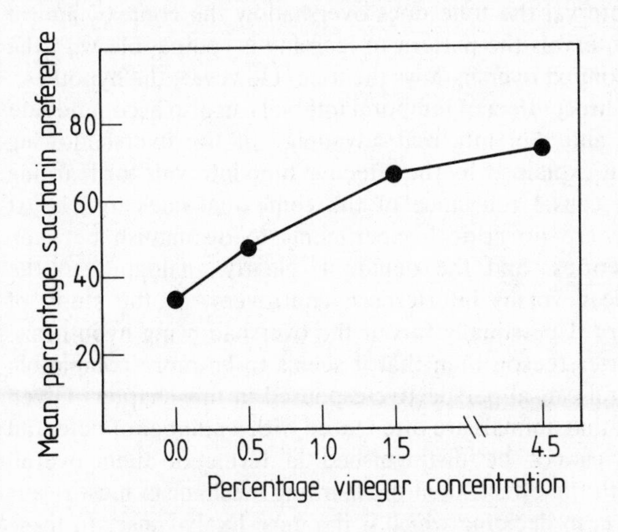

Fig. 18. The preference of rats for drinking from a spout delivering saccharin rather than one delivering a novel caffeine solution. Prior to this preference test all the rats had received in sequence access to saccharin and vinegar solutions followed by the induction of illness. The concentration of the vinegar solution was varied across the different groups. (After Revusky, 1971.)

interposed vinegar solution increased, the strength of the aversion to the saccharin decreased. So it appears that interposing a relevant and salient stimulus between exposure to saccharin and the induction of illness can reduce the amount learned about the saccharin → illness association. It is not unreasonable to argue that rats normally associate tastes and illness across such extended delays because the environment provides very few intervening relevant events to overshadow the target taste. By contrast, for tone → shock learning the environment typically contains a host of potentially relevant stimuli.

Where has this discussion led us? I pointed out that the general causal hypothesis offered no *a priori* reason why learning should be sensitive to the time interval between the constituent events as long as E2 did not occur before E1. However, experiments have shown us that it clearly is sensitive to this interval. We then considered two theories about why this might be so; one argued that E1 → E2 learning is directly sensitive to the time interval, whereas the other suggested that changes in this interval affected the degree to which contextual stimuli overshadow E1. When the amount learned about the context → E2 association is directly measured, we found that at the optimal interval the tone does overshadow the context, and at non-optimal intervals the pattern of learning is compatible with the idea that the context overshadows the tone. However, the hypothesis that there is a direct effect of temporal interval can also accommodate these results, and the only real advantage of the overshadowing theory is that it explains why the effective time intervals for learning vary with the causal relevance of the contextual cues to E2. At present, there are no critical experiments to distinguish between these two theories, and the debate is clearly analogous to the continuing decay versus interference controversy in the study of human memory. I personally favour the overshadowing hypothesis, but for no better reason than that it seems to be more compatible with the general causal perspective espoused in this chapter. Given that we accept that animals are often faced with a number of potential causes which cannot be distinguished in terms of their overall correlation with the effect, then the learning mechanism must resort to other criteria in deciding which is the most likely cause. In these circumstances, animals probably attribute the effect to the most salient and causally relevant E1. Furthermore this learning may occur at the expense of learning about the other potential causes. Similarly, we might argue that an animal will learn most about an event which is present at some optimal temporal relationship to the effect, and that such learning also detracts from the amount that animals learn about other potential causes.

Summary

In the previous chapter we discussed the types of relationships or associations between events which animals can learn. We saw that animals learn something about the association between two events,

E1 and E2, both when E1 is a potential cause of E2, or at least a detectable index of an underlying cause, and when E1 is a potential cause of the non-occurrence of E2. These two relationships we referred to as E1 → E2 and E1 → no E2 associations respectively, where E1 can be either a change in environmental stimulation or a component of the animal's own behaviour. When E1 is a stimulus, learning can be studied by observing a change in the animal's reactions to E1 in a classical conditioning procedure, whereas when E1 is an action learning is indexed by a change in the propensity of the animal to exhibit this behaviour. However, our discussion has been restricted primarily to classical conditioning for the technical reason that the experimenter has control over the frequency with which E1 occurs only in this paradigm. Such control is necessary if we are to discover exactly how the strength of learning is affected by arranging different relationships between E1 and E2.

There is good evidence that animals can learn about both E1 → E2 and E1 → no E2 associations. If E2 is a motivationally significant event for the animal, E1 → E2 learning usually leads to a direct change in the animal's reaction to E1. By contrast, in many situations E1 → no E2 learning remains behaviourally silent, and has to be revealed by running summation and retardation tests. In the present chapter we also considered the possibility that another form of behaviourally-silent learning could occur, namely that animals could actually learn that E1 and E2 were unrelated. We found evidence for two forms of such learning. In the first case, animals appeared to be able to learn that the occurrence of E2 was unrelated to a particular E1, whereas in the second case E2 was unrelated to the whole potential range of E1s. In this latter case the range of E1s was the animal's whole behavioural repertoire; this type of learning is referred to as learned helplessness.

Having considered the different types of learning, our attention was directed to the conditions under which these various forms of learning develop. For the animals to learn about an E1 → E2 association, there must be an overall positive correlation between the two events, or in other words the probability that E2 will occur in the presence of E1 or shortly after it must be greater than the probability that E2 will occur in the absence of E1. In order to understand why animals are sensitive to this overall correlation, one must appreciate that there are always more potential causes of E2 in the environment than just the target stimulus E1. Once this is realized, we can describe different correlation between E1 and E2 in terms of the pairings which occur between the other stimuli, the contextual cues, and E2 and between E2 and the compound of the contextual cues and E1. This analysis then allows us to understand why animals are sensitive to the overall correlation between two events in terms of the process of overshadowing and blocking. In turn, the problem of blocking was reduced to the question of why the amount that an animal learns from a series of pairings of E1 and E2 depends upon the extent to which the occurrence of E2 is surprising, or unpredicted, by other cues in the environment.

A similar analysis of the conditions for E1–no E2 learning showed that such learning occurs when the probability that E2 occurs in the presence of E1 is less than the probability of E2 in the absence of E1, or in other words when there is an overall negative correlation between the two events. This negative correlation can also be reduced to a sequence of pairings involving contextual cues, E1 and E2. For E1 → no E2 learning to occur, E1 has to be presented in the absence of E2, but in the presence of another stimulus which is independently an element in a stimulus → E2 association.

Finally we looked at two other conditions affecting learning. The amount that an animal learns about an E1 → E2 association depends upon how causally relevant the two events are. When two different E1s are paired with a particular E2, the animal often learns more about the relationship with one E1 than with the other. This effect depends upon the causal relevance of E1 to E2; when an appropriate E2 is employed, more is learned with the initially ineffective E1. The second condition concerned the temporal relationship between E1 and E2. The amount learned depends upon the temporal relationship between E1 and E2 with the optimal condition being one in which the onset of E1 occurs shortly before the onset of E2. Two hypotheses about the nature of this effect were considered. According to the first, varying the time interval has a direct effect on learning, whereas the second maintains that changes in this time interval increase the extent to which E1 can be overshadowed by contextual cues. The main advantage of this second idea is that it provides some account of why the effective intervals for learning should depend upon the type of events involved. Where the contextual cues are causally irrelevant to E2, then the effective interval with a relevant E1 should be long.

This discussion has not exhausted the various conditions which affect event learning. However, it has provided us with an adequate background against which to present an analysis of the associative representations set up during learning and the various theories about the nature of the mechanism underlying learning.

3 Associative representations

In the first chapter I outlined the three main topics in the study of simple associative learning: the conditions under which learning occurs, the nature of the internal representation encoding the learning experience, and finally the properties of the mechanisms producing learning. We have considered the first topic in the previous chapter, and we must now turn our attention to the nature of the representations set up during learning. The problem of specifying how human beings represent knowledge within the mind has become one of the central concerns of cognitive psychology over the last ten years or so, and a variety of systems has been developed (see Anderson, 1976). Many of these theories are concerned primarily with the relationship between knowledge structures and human linguistic capacities and, as a result, have employed channels of investigation which are not open to the student of infrahuman cognition. The theories of cognitive structures in animals, which we shall consider in this chapter, are much more primitive than their human counterparts, but this must not lead us to suppose that animal cognition is necessarily simple. The medium of language provides the student of human cognition with a rich data base that might reflect the structure of mental organization more directly than the purely behavioural capacities from which the animal psychologist must make his inferences. The relative poverty of our current theories of animal cognition probably reflects our inability to interpret the significance and meaning of an animal's actions, rather than limitations in the cognitive processes underlying its actions.

Declarative and procedural representations

In the brief discussion of internal representations of knowledge in chapter 1, I drew a distinction between two general forms of representation, the declarative form and the procedural form. According to the declarative model, knowledge is represented in a form which corresponds to a statement or proposition describing a relationship between events in the animal's world. This representation does not commit the animal to using the information for any particular function and assumes that a general set of processes can operate on these representations both to translate this knowledge into action and to integrate disparate, but relevant, items of information. The procedural model, on the other hand, assumes that the structure of the representation directly reflects the use to which the knowledge will be put in controlling the animal's behaviour.

I shall attempt to give these somewhat vague ideas substantive value by developing them in the context of a specific example. In a recent experiment Holland (1977) studied the behavioural changes that occurred when hungry rats were exposed to a tone → food relationship. This relationship was set up by occasionally presenting an 8-second tone and delivering food pellets into a receptacle or magazine immediately the tone ended. During the course of learning Holland observed a number of behavioural changes, one of which consisted of a developing tendency to approach the food magazine during the tone. One way of accounting for this particular behavioural change is to suppose that the learning experience sets up some sort of internal procedure for activating the motor programmes which execute the approach behaviour in the presence of the tone. As we currently know very little about the way in which such procedures might be represented, I shall attempt to capture their essential features by using the representational medium provided by natural English. Thus we might argue that the learning experience resulted in the formation of a procedure which was equivalent to the English instruction 'when the tone is on, approach the food magazine'. It cannot be emphasized too strongly that I do not intend to imply that the representational medium actually used by the rat is anything like natural English; rather the English instruction is given to illustrate the type of information and control encoded in such a procedural representation.

This procedural account assumes that the information is stored in a form which is closely related to the way the animal uses it. Furthermore, this form is assumed to be directly manifest in the animal's behaviour. By contrast, the information in a declarative representation is stored in a way that is not so strictly tied to use and can be retrieved or selected for a number of different functions. Thus in example we are considering the information acquired about the tone → food association might be stored in the form of a proposition equivalent to the English statement 'the tone causes the food'. The problem here is that the representation provides no direct account of why learning results in the particular behavioural change that we observe, in this case magazine approach. A declarative representation is simply a passive store of information which resides within the

Table 3. *Integration of associative representations*

Association	Procedural representation	Declarative representation
tone → food	'When the tone is on approach magazine'	'the tone causes food'
food → illness	'When food is present suppress eating'	'the food causes illness'
integration		'the tone causes illness'

animal's mental apparatus, and if such information is to control the animal's behaviour, some process must operate on this representation to translate it into procedures for governing the animal's behaviour. By omitting this translation stage, the procedural model provides a more simple and straightforward account to the behavioural changes observed during learning, and is clearly compatible with the behaviourists' idea that learning consists of a change in behaviour. However, this simplicity is bought at the cost of severely limiting the uses to which the information can be put. This restriction can be most easily appreciated by considering the ability of animals to integrate different items of information.

The integration of knowledge

Let us return to Holland's experiment in which hungry rats were exposed to a tone → food association and as a result developed a tendency to approach the magazine during the tone. Suppose we now devalue the food in a second stage by exposing the animals to a food → illness relationship until they refuse to eat this particular food when it is presented. These rats will now have learnt about two separate, but clearly relevant, relationships, the tone → food association and the food → illness association. The question is whether they are capable of integrating these two items of information, for such integration would tell us something about the form in which the knowledge about each association is represented.

Table 3 illustrates the possible English equivalent of the representations of each association according to the procedural and declarative models. The question of interest is whether the representation encoding each association can be sensibly integrated.

As stated, there is no straightforward inference rule for integrating

Table 4. *Design of the Holland and Straub (1979) experiment*

Group	Stage 1	Stage 2	Test
E	noise → food	food → LiCl	noise
C	noise → food	LiCl	noise

LiCl: lithium chloride

the two procedural representations, for they have no terms in common. By contrast, the two declarative representations have the 'food' term in common and this provides a basis for integration. An obvious integration gives the proposition 'the tone causes illness'. If rats can in fact perform such an integration, we should expect the behaviour controlled by the tone to change following exposure to the food → illness association.

Recently, Holland and Straub (1979) have performed an experiment which allows us to assess whether or not such integration occurs. The relevant aspects of their experimental design are outlined in table 4. In the first stage all the rats received a series of sessions containing a number of presentations of a 10-second noise stimulus each of which ended with the delivery of two food pellets. Figure 19A shows

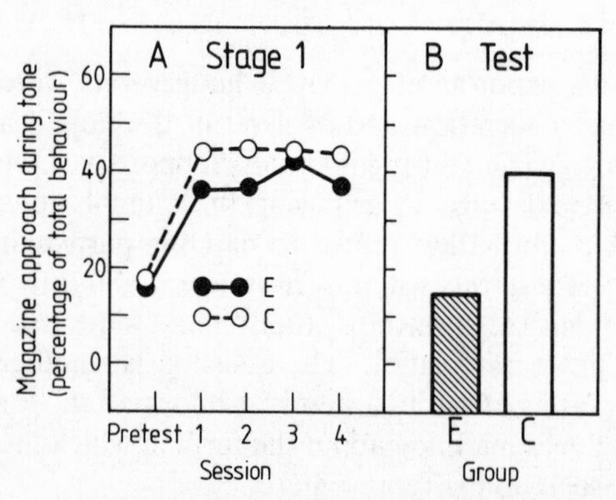

Fig. 19. Panel A: The acquisition of magazine approach during a tone which was paired with the delivery of food to the magazine. The strength of magazine approach is expressed by the percentage of all behaviours observed during the tone which consisted of approaching the magazine. Panel B: The strength of magazine approach after the food had been paired with the induction of illness in Group E but not Group C (see table 4). (After Holland & Straub, 1979.)

that magazine approach progressively became a dominant behaviour during the noise. In the second stage the rats in Group E received 50 pellets in their home cage. After 5 minutes during which they could consume as many of the pellets as they wished, the rats were injected with lithium chloride to induce illness. This procedure was repeated four times on different days and by the fourth food–lithium chloride pairing there was a dramatic decrease in the number of pellets the rats consumed, indicating that they learned about the food → illness association. The control group, Group C, simply received the injections of lithium chloride without the prior presentation of food. The devaluation procedure employed by Holland and Straub (1979) was in fact somewhat more complex. However, the above description captures the features that are necessary for our exposition.

In the final test stage the rats were returned to the experimental chamber where they received a number of presentations of the noise. If the animals in Group E were able to integrate the information acquired during exposure to the food → illness association with that acquired about the tone → food relationship, we might expect them to show a different pattern of behaviour from that exhibited at the end of Stage 1. Figure 19B indeed illustrates that Group E were less likely to approach the magazine during the noise after they had been exposed to the food → illness association. Group C, which had simply experienced the induction of illness, did not show such a change. The integration of knowledge demonstrated by this experiment requires that the animals store information about the associations in a representation that is sufficiently flexible and general to allow it to be integrated with other related representations. Clearly the declarative model outlined in table 2 is the most appropriate in this case.

Although there is little doubt that animals can integrate information about relevant event associations which have been learned at different times, it would be a mistake to conclude that such integration occurs automatically. That it does not can be illustrated by some further experiments conducted by Holland and Rescorla (1975a). Although basically similar to the Holland and Straub (1979) study, these experiments extend the chain of associations which the animal is required to integrate and are consequently somewhat more complex. The design of one of them is illustrated in table 5. Let us start out by considering the first two groups, Groups 2-E and 2-C. Initially the hungry rats in both groups were exposed to a light → food association which, in terms of our declarative model, should set up a representation corresponding to the proposition 'the light causes

Table 5. *Design of the Holland and Rescorla (1975a) experiment*

Group	Stage 1	Stage 2	Stage 3	Test
2-E	light → food	tone → light	food → rotation	tone
2-C	light → food	tone → light	rotation	tone
1-E	light → food	tone → food	food → rotation	tone
1-C	light → food	tone → food	rotation	tone

food'. In Stage 2 the animals experienced a second relationship, a tone → light association. Thus by the end of Stage 2 the rats should know about two separate, but related, relationships, the light → food and the tone → light associations. Given the results of the previous experiment, the animals should have been able to integrate information from these two associations to produce a representation corresponding to the proposition 'the tone causes food'.

If the animals do in fact perform this integration, the tone should have controlled a similar behavioural pattern after pairings with the light to that seen when the tone was associated with the food directly. To assess whether this was so, Holland and Rescorla included two further groups, Groups 1-E and 1-C. In the first stage these rats also experienced a light → food association, but in Stage 2 the tone was associated with the food directly rather than with the light as for Groups 2-E and 2-C (see table 5). If the animals in Groups 2-E and 2-C integrated the information about the light → food and tone → light associations, the tone should have elicited the same response pattern in all the groups during Stage 2. Holland and Rescorla in fact only studied a very gross measure of the animals' behaviour during the tone by simply recording the rats' total activity. The way in which this total activity measure changed during Stage 2 is illustrated in figure 20A. All groups showed an essentially comparable increase in general activity. Although Holland and Rescorla (1975a) did not analyse the changes in specific response patterns, such as magazine approach, underlying these gross activity changes, Holland (1977) has subsequently done so. Essentially the same pattern was observed when the tone was paired with food directly as when it was associated with a light that had been previously paired with food. In the terminology of conditioning and reinforcement, the activity changes seen in Groups 2-E and 2-C during Stage 2 are referred to as higher-order conditioning (see p. 42) in contrast to the

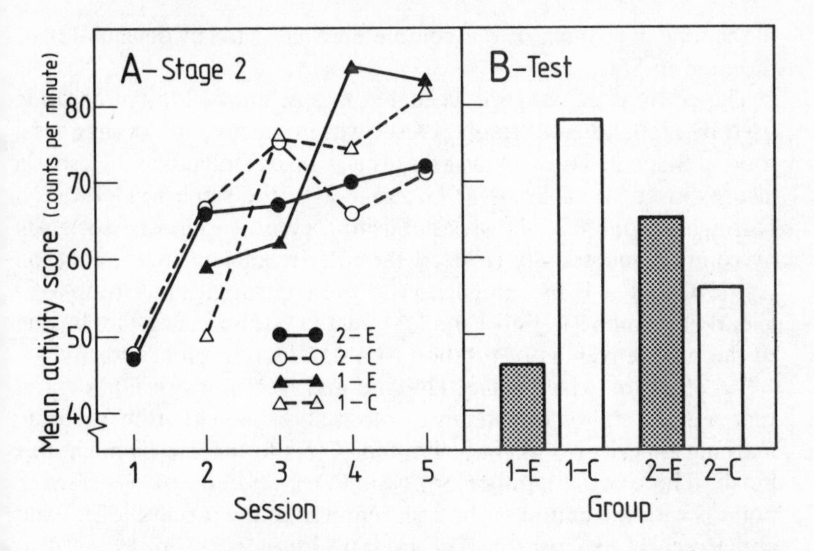

Fig. 20. Panel A: The changes in general activity produced by a tone directly paired with the delivery of food in Groups 1-E and 1-C and a tone paired with a light which had itself been previously associated with food in Groups 2-E and 2-C (see table 5). Panel B: The general activity produced by the tone on test after the food had been paired with the induction of illness in Groups 1-E and 2-E but not in Groups 1-C and 2-C (see table 5). (After Holland & Rescorla, 1975a.)

first-order conditioning exhibited by Groups 1-E and 1-C.

So far all the results are congruent with the idea that animals can integrate information across a number of associations. Holland and Rescorla then went on to test this integrative power further by adding a third association to the chain. In Stage 3 Group 2-E experienced a food → illness association, which was established by rotating the rats after they had eaten the food rather than by injecting a mild poison. Pairing the intake of a particular food with rotation has been shown to establish an aversion akin to that seen with injection of lithium chloride. So in Stage 3 Group 2-E received a series of food–rotation pairings, whereas the control group, Group 2-C, simply experienced the rotation in the absence of the food. Animals in Group 2-E now had the opportunity to integrate information from three associations, the light → food, the tone → light, and the food → illness association, to arrive at the proposition that 'the tone causes illness'. If such integration occurs we should expect the reactions of Group 2-E to the tone on test to be different from those of Group 2-C. The results of this test stage are shown in figure 20B. The behaviour of the animals to the tone in Group 2-E was completely unaffected by devaluation of the food in Stage 3.

The obvious explanation is that rotation, unlike lithium chloride used by Holland and Straub (1979), was ineffective in devaluing the food in Stage 3. To check whether this was so, Holland and Rescorla also exposed the animals in Group 1-E to the same food–rotation pairings. As figure 20 shows, establishing a food → illness association by rotation successfully reduced the activity elicited by the tone on test in Group 1-E by comparison to the control group, Group 1-C. Clearly the failure to find a similar effect in Group 2-E cannot be due to the ineffectiveness of rotation as a devaluation procedure.

Taken at their face value, Holland and Rescorla's results suggest that a rat's ability to integrate information gained from separate learning experiences might be limited. The question is why might this be so. There are a number of possibilities, and an obvious starting point is with the nature of the task required of the rats in the Holland and Rescorla experiment. The rats in Group 2-E were exposed to three associations; tone → light, light → food, and food → illness. In order to show an appropriate behaviour change on test they were required to integrate the first, the tone → light association, with the third, the food → illness association. However, as these two associations do not have any terms or events in common, such integration would have to be mediated via the second relationship, a light → food association. Perhaps the form of the representations set up by these learning experiences and nature of the processes operating upon them do not allow for such mediated integration. We can investigate this possibility exposing the animals to three associations which all have an event in common. Consider the case in which the rats are first exposed to light → food and tone → light associations, as in the Holland and Rescorla (1975a) experiment, but then experience a light → no food relationship. The design of such an experiment is outlined in table 6. We already know that the rats will be able to integrate the light → food and the tone → light associations presented in the first two stages; the question is whether they will then be able to do the same for the tone → light association and the light → no food relationship presented in the third stage. If the animals can integrate information from any two associations containing an event in common, such integration should occur and would be manifest in a loss of the response pattern established to the tone in Stage 2.

Holland and Rescorla (1975b) have performed an experiment exactly like that outlined in table 6. The training during the first two

Table 6. *Design of the Holland and Rescorla (1975b) experiment*

Group	Stage 1	Stage 2	Stage 3	Test
E	light → food	tone → light	light → no food	tone
C	light → food	tone → light		tone

stages was the same as in the previous experiment. In Stage 3 Group E simply received presentations of the light alone until the responses established to it in Stage 1 completely disappeared, whereas the control group, Group C, received no light presentations during this third stage. The results were clear cut; the pattern of responding, established to the tone during Stage 2, was elicited to the same extent in Groups E and C by test presentations of the tone in the final stage. In this case the failure of integration cannot be due to the fact that the two target associations, the tone → light and the light → no food relationships, do not contain an event in common. It should be noted, however, that failures of integration do not always occur with the procedure outlined in table 6. Both Rashotte, Griffin and Sisk (1977), and Leyland (1977) have reported integration using autoshaping of the pigeon's peck response.

At present we do not really know why successful integration occurs in some cases but not others. In the procedure studied by Holland and Rescorla the problem seems to revolve around the tone → light association, for they have been consistently unsuccessful in demonstrating integration involving this association once it has been established. One possibility is that exposure to this association sets up a procedural representation of the form specified in table 3. The integrative processes would be unable to use a representation of this form. An alternative, which we shall consider in more detail in the next sections, is that the learning experience forms a representation which is intrinsically declarative in form but does not encode the constituent events, such as the light, in the same way as for other associations involving the light. Unless the integrative processes can identify the 'light' that occurs in the tone → light association as being the same event as the 'light' in the light → food and light → no food relationships, there is no basis for integration. This argument may seem somewhat obscure at present, but it will be developed in detail in the next sections.

I started this section by pointing out that there are two general classes of knowledge representation, a procedural form and a declarative form. The declarative form provides a more suitable basis for the integrative processes and the fact that animals are clearly capable of performing such integration in certain cases suggests that they can encode information in declarative form. By a similar argument, the failures of integration, which we have discussed, may be seen as evidence that procedural representations are also employed. The capacity to integrate, however, must not be taken as a definitive test of the form of the representation, and in fact there is probably no such definitive test. The representations, given as illustrations of the procedural and declarative forms in table 2, are very simple and restricted ones, and once we allow the possibility that more complex and powerful forms, be they procedural or declarative, may be used, there is no limit to the mental operations which can be performed with both types of representation. At present, the only argument we can use to evaluate the two models is to ask whether a particular cognitive ability, such as the integration of information, is more compatible with one form of representation rather than the other. The success or failure of integration does not conclusively determine the nature of the underlying encoding but is at least suggestive of it.

Knowledge–action translation

So far we have only discussed the nature of representations in the broadest terms by drawing an analogy between English statements and commands and the possible forms of encoding used by animals. We must now attempt to specify the nature of these representations more precisely. As I pointed out earlier, we cannot investigate their nature by asking infrahuman animals questions, and so we must rely on a study of the way in which animals use their acquired knowledge to control their behaviour. Up to now we have paid little attention to the actual form of the behavioural changes observed during learning, and have been content to use such changes simply as an indicator of whether or not learning has occurred. However, the form of these behavioural changes must be determined, at least in part, by the way in which the knowledge is stored and the nature of the processes that operate on these representations to translate them into action.

One feature of the behavioural changes we observe during E1 → E2 learning tends to stand out above all others; in the case where E1 is some environmental stimulus, the form of the response elicited by E1 is similar to the form of the response normally elicited by E2. Holland and Straub (1979), for instance, found that their rats came to approach the food magazine during a tone which was paired with delivery of food into the magazine. Such approach responses, of course, are similar to those elicited by E2 in this situation, the stimuli and events accompanying the delivery of food pellets into the magazine. I could carry on enumerating examples of this similarity. When a rabbit experiences pairings of a tone and a mild shock or air puff to the region of the eye, the tone elicits a blink of the eyelid which is very similar to that elicited by the shock or air puff. In Pavlov's classical conditioning procedure during which a dog experiences pairing of a tone and food, the tone, like the food, elicits salivation. Pavlov was the first to note the similarity of the behavioural patterns elicited by E1 and E2 during learning about an E1 → E2 relationship and suggested that, as a result of this learning, E1 becomes a substitute or surrogate for E2.

An elegant example of the operation of this so-called principle of 'stimulus substitution' comes from a study by Jenkins and Moore (1973). In the first chapter we considered an experiment by Wasserman *et al.* (1974), in which pigeons were exposed to pairings of the illumination of a small disc or key in the wall of their chamber for 10 seconds and the presentation of food. As a result of experiencing this light → food relationship, the pigeons came to approach the key as soon as it was illuminated. However, not only will the pigeons approach this key, but they will also start to peck it. This pecking response seems like yet another example of stimulus substitution in that the behaviour elicited by E2, in this case pecking the food, comes to be elicited by E1, the lighted key. Given this basic similarity, Jenkins and Moore attempted to find out whether the form of the response elicited by E1 was really determined by the form of the behaviour controlled by E2.

They started out by noting that the pigeon's consummatory responses to food and water are different. When eating grain the pigeon makes short, sharp, pecking movements during which the beak is open at the moment of contact and the eyelids almost closed. By contrast, when drinking, the bird just places its beak into the water, slightly open, and performs swallowing movements with its eyes fully open. Jenkins and Moore argued that if the principle of stimulus substitution is controlling the form of the response directed at the lighted key, different types of responses should be observed when the key is paired with food and water. The pigeons were trained in an operant chamber with two keys while they were both hungry and thirsty. Illumination of the right-hand key for 6 seconds was followed by the presentation of grain in a food magazine, whereas the illumination of the left-hand key was followed by the presentation of water. These two types of trial were presented in a random order. With training the birds came to approach and contact both keys with their beaks whenever the keys were illuminated. The critical observation, however, was that the form of the response directed at the right-hand key, the one paired with food, resembled the consummatory response for food, whereas the response directed at the left-hand key resembled the drinking pattern. In other words, the pigeons appeared to attempt to eat the key associated with food, and drink the one associated with water.

Obviously any account of the representation set up by this type of learning experience must encompass the principle of stimulus substitution, and there are two traditional models in the field of animal learning which attempt to do just this, one declarative in form and the other procedural. We shall consider the declarative model first, which was initially proposed by Pavlov (1927) himself and subsequently developed by Konorski (1948; Dickinson & Boakes, 1979). A schematic representation of this model is illustrated in figure 21. In

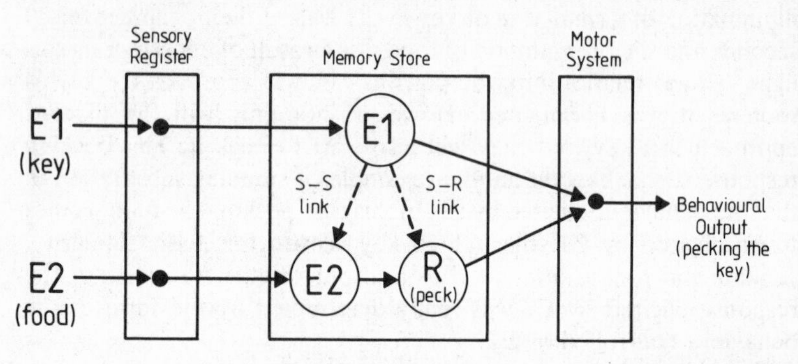

Fig. 21. A schematic illustration of the possible excitatory-link associative structures set up by exposure to an E1 → E2 association. The S–S link corresponds to a declarative representation and the S–R link to a procedural representation. Presenting E1 activates an internal representation of E1 in the animal's memory store which in turn activates either a representation of E2 via an S–S link or the representation of the response normally elicited by E2 via an S–R link. The observed behaviour is an amalgam of the influences of the aroused E1 and response representations.

describing the model let us start with the reactions of a hungry pigeon to food. The presentation of food is detected by the bird's sensory and perceptual systems, which we shall refer to as a sensory register, and the output of this system is assumed to activate or excite some internal representation of the food in the animal's memory store. The activation of this event representation will in turn excite the representation of the appropriate response, such as pecking, via either innately determined or previously acquired connections or links. An aroused response representation will lead to the performance of the appropriate behaviour via the motor system. According to this model, learning about a key → food association consists of the formation of a link between the internal representation of the food and some representation of the lighted key. After the formation of this link presentation of the illuminated key will activate its representation which in turn will arouse the representation of the food via the acquired link thus producing the pecking response. Of necessity the response elicited by the key must resemble that produced by the food itself.

This associative representation is essentially declarative in form. Like the English statement 'the key causes food', it consists of three components; it has a constituent representing the key, one representing the food, and a connection representing the relationship between these events. The properties of this connection, however, are very limited by comparison with the relational terms which can be expressed in a system like the English language. It is simply an excitatory link which has no other property than that of transmitting excitation from one event representation to another. Whereas in English we can state a large number of different types of relationship by employing different terms, in this model there is only one term (or possibly two, as we shall see later) to represent an association. This restriction imposes severe limits on the model's capacity to store different types of knowledge. The concept of the excitatory link can be embodied in a number of different ways. As developed by Konorski, for instance, the excitatory link was viewed as effectively a neural connection between the brain system activated by E1 (the key) and E2 (the food). However, there is no need to express the concept in terms of such hypothetical neurophysiological mechanisms, and we could equally well view an excitatory link as a retrieval process which permits the presentation of E1 to retrieve from the animal's memory store information about E2 (Wagner, 1978). In cases where E1 is a stimulus, as in Pavlovian conditioning, this type of declarative excitatory link representation is usually referred to as a stimulus–stimulus, or S–S, representation. The process which operates on this S–S representation to translate knowledge into action is correspondingly simple; it just consists of activating the representation of E1, either by actually presenting E1 itself or by arousing it via some other excitatory link to which it is already connected. As soon as the E1 representation is activated, it will immediately transmit this activation to the E2 representation and thus to the motor system. The process for integrating knowledge is also simple. If we expose an animal to a stimulus → food association and then to a food → illness association, excitatory links will be formed between the representations of the stimulus and the food and then between those of the food and illness. Consequently, we shall have a two-link chain, whereby exciting the stimulus representation will now activate the illness representation via that of the food.

We can also construct a simple procedural model in terms of excitatory links. All we have to do is to assume that exposure to an E1 → E2 relationship results in the formation of an excitatory link from the E1 representation directly to the response representation activated by E2. Such a connection is also illustrated in figure 21. This model corresponds to the well-known stimulus–response, or S–R, theory developed by the neobehaviourists, such as Hull and Guthrie. As this S–R representation does not contain a constituent corresponding to E2, any form of integration which involves a change in the significance of E2 cannot be handled by this procedural theory. As we have seen, in certain circumstances animals appear to be unable to integrate information from different associations, and in these cases this procedural model is clearly a strong candidate for the underlying representation.

The usual objection to these models is to point out that the form of the response elicited by E1 does not always bear a marked similarity to that elicited by E2. However, such divergences must be inevitable in certain cases. Suppose that instead of experiencing the pairing of a spatially localized stimulus, such as a lighted key, with food, the pigeons were exposed to a relationship between food and a diffuse stimulus, such as a tone (Schwartz, 1973) or a change in the colour of the ambient illumination of the chamber (Blanchard & Honig, 1976). In these circumstances we could not possibly expect to observe a response closely resembling the pattern elicited by the delivery of food; the pigeon just cannot peck at a diffuse tone or light. However, there is no reason to believe the underlying associative representation is different in the two cases. In fact there are good empirical reasons for believing that they are similar. Blanchard and Honig (1976) exposed one group of pigeons to pairings of a diffuse red light and the delivery of food in the first stage of a two-stage blocking experiment (see p. 46). Exposure to this association did not lead to the development of any observable pecking behaviour. However, if in the second stage the pigeons were exposed to pairings of a lighted key and food in the presence of this red light, the frequency with which the animals pecked this key was much lower than in a control group for whom the red light had not been previously associated with food. In other words, experience with a red light → food association blocked learning about the key light → food relationship although the two associations did not produce the same overt behavioural changes. The presence of this blocking effect indicates that similar representations are set up by the two associations even if they are not similarly manifest in overt behaviour.

One way of accounting for the effect of the nature of E1 on the form of the observed behaviour is to asume that this behaviour reflects some form of amalgam of the influence of E1 representation with that of the E2 and its associated response representations on the motor system (see figure 21). In order to observe true stimulus substitution, E1 must provide the appropriate stimulus support for the type of behaviour normally elicited by E2. There are numerous examples in which the type of behavioural change observed during E1 → E2 learning is not only a function of E2, but also reflects the significance of E1 for the animal. One of the most dramatic comes from a study by Gustavson, Kelly, Sweeney, and Garcia (1976) in which wolves were made ill by consuming poisoned meat wrapped in sheep skin. Subsequently when presented with a live sheep, not only did the wolves refrain from eating it, they did not even attack it. In fact after a few minutes of interaction with the sheep, the wolves actually exhibited a pattern of submissive behaviour which would normally be elicited by a dominant member of their own species. In this example it is clear that the way in which the wolves' knowledge about the food → illness association was translated into action was radically affected by the fact that E1, the food, was presented in an animate form.

The principle of stimulus substitution basically asserts that exposure to an E1 → E2 association endows E1 with the properties of E2. We have already seen that E1 acquires the capacity to elicit responses similar to those produced by E2. However, many events of the type we have been considering, such as the delivery of food to a hungry animal, not only have the capacity to elicit certain response patterns, but also have motivational or affective value for the animal. Certain events are attractive, whereas others are aversive or noxious.

An obvious question is whether E1 acquires the same affective value as E2 following exposure to an E1 → E2 association. If E2 is an attractive event for the animal, E1 should also be attractive. Correspondingly E1 should be aversive if E2 is noxious. There is little doubt that this is so. For instance, Hyde (1976) exposed hungry rats to a number of pairings of a 3-second tone and the delivery of food. Food is clearly an attractive stimulus for hungry rats in the sense that they will perform some activity or response, such as pressing a lever, which produces food. If the tone, as result of being paired with the food, acquires the same motivational significance, the rats should also be prepared to perform an action which produces the tone. Hyde (1976) tested whether this was so by presenting the rats with a lever in the second stage in the absence of any food. Each lever-press produced the tone. A control group received exactly the same training except that during the first stage the tone and the food were presented randomly with respect to each other. In chapter 2 we saw that this random condition represents the appropriate control against which to assess any effect due to associative learning. If exposure to a tone → food association endows the tone with the same motivational and rewarding properties as the food, we should expect the animals in the paired condition to respond more rapidly on the lever than those in the random condition. This is just what happened.

In conditioning-reinforcement terminology, the tone is referred to as a secondary or conditioned reinforcer in the paired condition. By strengthening the lever-press response, the tone acts as an instrumental positive reinforcer, and this property is acquired through the classical or Pavlovian conditioning procedure in the first stage. Of course, if the tone had been paired with an aversive stimulus, such as a shock, in the first stage, it would have suppressed lever-pressing in the second stage, and thus acted as a conditioned punisher. This type of experiment provides convincing evidence that exposure to an E1 → E2 association not only enables E1 to elicit similar responses to E2, but also endows E1 with the same motivational significance as E2.

Variations in event encoding

So far I have assumed that exposure to a simple E1 → E2 relationship sets up a single underlying representation of this association. Such an assumption, however, may simply reflect the tendency of experimenters to study learning by looking at changes in a single response system. Exposure to an E1 → E2 relationship usually produces a diverse range of behavioural changes, and if we restrict ourselves to measuring only a single response system, we may very well fail to detect a rich pattern of representations formed by a simple learning experience.

When more than one response system has been systematically studied, the pattern of behavioural change has often indicated the presence of more than one underlying representation. For example, Vandercar and Schneiderman (1967) exposed rabbits to pairings of a tone as E1 and a mild eyeshock as E2 and then measured learning by recording whether the tone produced a change in heart rate and an eyeblink. When the interval between the onset of the tone and the delivery of the eyeshock was only 0.75 seconds, the tone gradually acquired the ability to elicit both responses. By contrast, figure 22

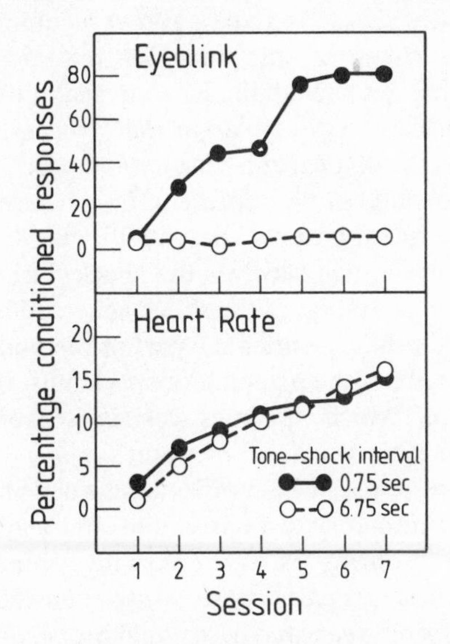

Fig. 22. The percentage of tone presentations on which a tone elicited an eyeblink (top panel) and a change in heart rate (bottom panel) during sessions in which the tone was paired with the eyeshock. For different groups of rabbits the interval between the onset of the tone and the delivery of the shock was either 0.75 or 6.75 seconds. (After Vandercar & Schneiderman, 1967.)

shows that when the tone–shock interval was 6.75 seconds learning occurred just as readily as at short intervals according to the heart-rate measure, but completely failed to occur according to the eyeblink measure.

What are we to make of such results? One explanation could be that the relative independence of the eyeblink and heart-rate changes reflects the fact that exposure to the tone → shock association can set up two representations, one mediating the change in the blink response and the other alterations in heart rate. At short tone–shock intervals both representations are formed, whereas, for some reason or other, at long intervals only the representation producing the change in heart rate is set up. Konorski (1967) was the first to suggest that a simple learning experience might lead to the formation of two relatively independent representations. He noticed that the types of response change produced by a simple E1 → E2 association appeared to fall into two rough categories. First there were those responses which seemed to accord with the strict principle of stimulus substitution in that their form closely matched that of the response elicited by E2. The pigeon's key peck and the rabbit's eyeblink response are examples in this category. In order for this close matching to occur, the information encoded in the E2 representation, activated by presenting E1, must be very similar to that processed when E2 itself is presented. Specifically, certain sensory and perceptual properties of E2 must be encoded in the representation aroused by E1. For instance, in the case of rabbit eyeblink conditioning the tone only elicits a blink of the eye which receives the shock during training. This means that the representation of the shock activated by the tone must contain information that specifies the part of the body to which the shock is applied. Similarly the pigeon's key peck must be based on an E2 representation which specifies certain sensory properties of E2, in this case whether it is food or water.

Konorski referred to learned responses which reflected the specific sensory properties of E2 as consummatory responses, and contrasted them with a second class observed during E1 → E2 learning, preparatory responses. The form of these preparatory responses does not appear to be closely tied to the specific sensory-perceptual properties of the E2, but rather reflects its generally affective or motivational value for the animal. We have already considered a number of such preparatory responses. When a lighted key is paired with food, a pigeon tends to approach the key, as well as pecking at it (Wasserman *et al.*, 1974). This approach behaviour does not reflect any specific properties of the food, and would probably be observed if the key was paired with any other attractive stimulus for the pigeon, such as water or a mate. In this sense the preparatory approach response just reflects the general affective value of the food, namely that it is an attractive event, and can be contrasted with the pecking behaviour which is determined by the specific sensory-perceptual properties of the food. The rabbit's heart rate changes, observed during pairings of the tone and eyeshock, are probably also preparatory in nature in that similar changes might well be observed if the tone was paired with another E2 of the same motivational significance and value as the eyeshock. This other E2, however, would not produce the same consummatory response, the eyeblink.

If we accept this preparatory–consummatory distinction, Vandercar and Schneiderman's (1967) experiment suggests that the representations mediating changes in the consummatory eyeblink and preparatory heart-rate response systems are relatively independent. Remember that with a short tone–shock interval changes in both eyeblink frequency and heart rate are observed, whereas with a long interval only the change in heart rate occurs. This pattern could be explained by assuming that two independent representations are set up during learning, one mediating the heart-rate change and the other the eyeblink response. During learning, presenting the eyeshock activates two independent E2 representations, one encoding the sensory-perceptual properties of the eyeshock and the other its affective value. Exposure to the tone → shock association with a short interval leads to the formation of independent excitatory links between the representation of the tone and both the sensory-perceptual and affective representations. By contrast, when the tone–shock interval is long only the excitatory link between the tone representation and the affective representation of the shock will be formed, and thus only the heart-rate change observed. Konorski did not specify why preparatory learning is more resilient to increases in the E1–E2 interval.

It is important to realize the full implications of this type of model. On the most general level, it argues that the way in which information about E2 is encoded in a learned representation can vary with the conditions of learning. When the tone–shock interval is short, the rabbit essentially has two items of knowledge corresponding to the statements 'the tone causes a shock to the left eye' and 'the tone causes a mildly aversive event'. When the interval is long, however, only the second representation is set up, so that the animal knows that the tone causes something unpleasant, but does not know exactly what it is. It is as though in the first case presenting the tone both retrieves an image of a particular stimulus applied to one eye and makes the animal feel mildly anxious, whereas in the second case the tone simply makes the animal anxious without informing it of the reason for the feeling.

But is it really true that a learning experience can set up a representation which only encodes whether E2 is attractive or aversive, but carries no information about the specific nature of the event? If this was so, we should be able to take two events which are qualitatively very different on a sensory-perceptual level, but have

Table 7. *Design of the Bakal, Johnson and Rescorla (1974) experiment*

Group	Stage 1	Stage 2	Test	Suppression ratio
C		Light + Tone → Klaxon	Light	0.08
K	Tone → Klaxon	Light + Tone → Klaxon	Light	0.26
S	Tone → Shock	Light + Tone → Klaxon	Light	0.29

similar affective values, and by presenting them as E2 in an E1 → E2 association with a long E1–E2 interval produce the same behavioural changes. This seems to happen. For example, Bakal, Johnson and Rescorla (1974) paired a 30-second compound stimulus consisting of a light and a tone with either a foot shock or a klaxon which was sufficiently loud to be aversive. In both cases it was found that the light–tone stimulus acquired the ability to suppress lever-pressing by rats for food. Thus it appears that two very different events in terms of their sensory properties, a klaxon and a foot shock, will act as effective E2s to produce the same preparatory response change. The obvious implication is that the representations underlying the response changes are similar in both cases. In fact Bakal *et al.* (1974) produced some even more compelling evidence that the event representations set up by the klaxon and shock were similar by showing that learning about a stimulus → shock relationship would subsequently block learning about a stimulus → klaxon association. The design of their experiment, outlined in table 7, employed the familiar two-stage blocking design. In the second stage all rats received a series of presentations of a 30-second compound of a tone and light, each of which terminated with a 2-second presentation of the klaxon. The amount the animals had learned about the light → klaxon association in this second stage was measured by presenting the light alone in a final test stage and seeing by how much its presentation suppressed lever-pressing for food. As usual the group differed in the training schedule they received during the first stage. Group K experienced a series of trials in which the tone was paired with the klaxon. For this group learning about the light → klaxon relationship in Stage 2 should have been blocked by the pretraining to the tone in Stage 1. The final column of table 7 shows the suppression ratios produced by the light on test. Remembering that lower suppression ratios indicate more learning, we can see that Group K learned less about the light → klaxon association in the second stage than a control group, Group C, which was not pretrained to the tone in the first stage. This is the standard blocking effect.

In our discussion of the blocking effect in chapter 2, we argued that for an animal to learn about an E1 → E2 association the occurrence of E2 must be surprising or unexpected. Blocking occurs because pretraining to one of the elements in the first stage ensures that it is a good predictor of E2, so that when E2 is presented in the second stage it is not surprising and hence fails either to initiate or maintain learning. Given this perspective, let us see whether we should expect blocking in the critical group of interest in the Bakal *et al.* (1974) experiment, Group S. This group received the same training as the others in Stage 2, but in Stage 1 these rats experienced pairings of the tone with shock. The question is whether learning about a tone → shock association will subsequently block learning about the light → klaxon relationship. In terms of their sensory-perceptual properties the shock and the klaxon are very different events. Consequently, if exposure to these relationships sets up representations of the form 'the tone causes the shock' and 'the light causes the klaxon', we should expect the occurrence of the klaxon in Stage 2 to have been surprising and hence produce good learning about the light → klaxon relationship. On the other hand, if this type of learning procedure forms a representation which encodes predominantly information about the affective value of the events, blocking might be expected. Under these circumstances, the tone → shock association would result in a representation of the form 'the tone causes an aversive event', and as the klaxon is also a mildly aversive event, the expectation based on this representation would be fully confirmed by the delivery of the klaxon in Stage 2. In terms of the affective values of the events, the occurrence of the klaxon in Stage 2 would be fully predicted by the tone, and hence blocking should be observed. Table 7 shows that such blocking did in fact occur; pretraining with a tone → shock association in Group S blocked learning about the light → klaxon association just as much as pretraining with the tone → klaxon association in Group K.

The idea that the way in which events are encoded during learning can vary has received considerable attention in studies of human learning and memory. For instance, Craik and Lockhart (1972) have argued that verbal material can be processed to different 'depths' depending upon the manner in which the learning task is presented to the subject, and it is not unreasonable to assume that a similar variation in encoding might occur with infrahuman animals and non-verbal material. In cases where we observe strict stimulus substitution in certain response systems, such as rabbit eyeblink conditioning and pigeon autoshaping, information about the specific sensory and perceptual properties of E2 must be encoded by the learned representation. However, the transreinforcer blocking effect we have just considered suggests that under certain circumstances the representation set up by the learning experience only encodes information about the general affective or motivational properties of E2. It is not clear which features of the learning experience determine the type of encoding observed in animals, and at present we can do little more than endorse Konorski's observation that increasing the E1–E2 interval makes it more likely that we shall observe response changes which reflect solely the affective or motivational properties of E2.

In the previous section on the integration of knowledge, I suggested that the way in which animals can 'put together' or integrate information from independent, but relevant, associations tells us something about the form in which information about these associations is encoded. It was argued that evidence for integration points to an underlying representation with a declarative form. It is obvious, however, that the ability to perform such integration must be a function, not only of the overall form of the representation, but also of the way in which information about the constituent events is encoded. To illustrate this point let us reconsider the experiments by Holland and Rescorla (1975a,b), illustrated in tables 5 and 6. Recall that when rats were required to integrate information from two relationships, a tone → food and a food → illness association, they appeared to be perfectly capable of doing so. Exposure to the food → illness association reduced the behavioural change brought about initially by exposure to the tone → food relationship. This we took as evidence that these associations were encoded in a declarative form. Holland and Rescorla (1975 a,b) then went on to show that rats did not appear to integrate other types of relevant associations. If the rats were first exposed to a light → food association and then to a tone → light relationship, the representation set up by this latter association appeared to be resistant to integration. Devaluing either the light by exposure to a light → no food association or the food by exposure to a food → illness association left the strength of the response elicited by the tone unchanged. One interpretation of this finding is that exposure to the tone → light relationship sets up a representation whose form is incompatible with the integrative processes; the simple procedural form we considered is one example.

There is, however, a weakness in this argument. Once we accept that the way in which information about the constituent events is encoded within the representation can vary, the integration test becomes a less incisive criterion for distinguishing between procedural and declarative forms. When Holland and Rescorla presented their rats with a tone → light association at least three features of the light could have been encoded within a declarative representation of this relationship. The E2 representation could have contained information about the sensory and perceptual properties of the light itself, about the sensory and perceptual properties of the food which had been previously associated with the light, or about the general affective or motivational properties of the food or light. If the sensory-perceptual properties of either the light or the food had been encoded, devaluing at least one of these stimuli should have been effective in changing the animals' response to the tone. However, if the E2 representation encoded only the general affective values of the food and light, giving for example a representation of the form 'the tone causes an attractive event', there is no way in which a change in the value of the particular food or light used in these experiments could be meaningfully integrated with this representation.

Once we allow that only partial information about an event may be encoded, the integration test no longer provides a criterion for deciding between the procedural and declarative forms. Failures to integrate associations containing a common event may arise both when one of the representations contains no constituent element referring to that event, as in a procedural form, and when the way in which the event is encoded in the two representations differs. In fact the very notion of variations in event encoding can render the whole distinction between declarative and procedural forms somewhat blurred. For instance, I suggested that a declarative encoding may take the form 'the tone causes an attractive event'; however, the labelling of an event as attractive could be regarded as a response to a particular stimulus so that the representation could take a procedural form, such as 'when the tone is on, respond with the sensation of pleasure'. It is far from clear that these two are really distinguishable.

E1 → no E2 associations

In chapter 1, I described the two main types of relationship that animals learn about, E1 → E2 associations and E1 → no E2 associations. The discussion so far has been directed at elucidating the representations set up by exposure to E1 → E2 relationships, but now we must attempt a parallel analysis for E1 → no E2 learning. Such an analysis has provided an enduring problem for students of animal cognition, and it is far from clear whether we have as yet any satisfactory answers.

By analogy with E1 → E2 learning, the most straightforward account would be one in which exposure to an E1 → no E2 association sets up a representation of the form 'E1 causes the omission of E2'. The central problem with this idea concerns the way in which the concept of 'the omission of' is embodied in terms of a structure consisting of event representations in the memory store and links between these representations. There have been basically two ideas about the nature of such a structure, both primarily developed by Konorski (1948, 1967; Rescorla, 1979). The first assumes that the concept 'the omission of' is represented by the nature of the link itself. Thus exposure to an E1 → no E2 association would set up a representation in which a particular type of link, an inhibitory link, is formed between the representations of E1 and E2. The second account argues that there is a specific event representation which encodes the event 'the omission of E2'. According to this model the E1 → no E2 association is represented by an excitatory link between the E1 representation and a representation encoding an event, namely 'the omission of E2'. Thus the difference between the representation set up by E1 → E2 and E1 → no E2 associations is in the nature of the link according to the first model, but in the nature of the terminal event representation according to the second.

To adjudicate between these two models we shall have to look at the behavioural properties of E1 after exposure to an E1 → no E2 relationship. In our initial discussion of E1 → no E2 learning in the first chapter, we noted such learning is often behaviourally silent in that the overt responses elicited by E1 remain unchanged. However, we can reveal that some learning about the association has taken place by looking at the capacity of E1 to either affect subsequent learning or modulate the animal's responsiveness to other stimuli. Such effects were illustrated by a conditioned-suppression experiment reported by Rescorla (1969a). In the first stage rats experienced a negative correlation between a tone as E1 and a shock as E2. This negative correlation ensured that the shock never occurred either during the tone or shortly after it and thus established a tone → no shock relationship. When the tone was subsequently presented while the rats were pressing a lever for food, the rate of responding remained unchanged. Retardation and summation tests, however, were used to show that the rats had in fact learned about the tone → no shock association. In the summation test, for instance, the rats subsequently learned about a light → shock association so that the presentation of the light severly suppressed responding. It was then found that presenting the tone in compound with the light reduced the level of suppression. The tone, although without obvious direct behavioural effect itself, appeared to be capable of attenuating or inhibiting the learned response elicited by the light.

This inhibitory capacity of the tone can be directly explained by the inhibitory-link model. Exposure to the tone → no shock association will result in the formation of an inhibitory link between the tone representation and the shock representation. Consequently presentation of the tone will produce an inhibitory influence on the shock representation making it less responsive to an excitatory influence. Thus, when the light is presented in conjunction with tone, the degree to which it activates the shock representation will be reduced by the inhibitory influence of the tone.

Although such as account provides an elegant explanation of inhibitory properties, it has difficulty with the case in which E1 → no E2 learning does more than just endow E1 with an inhibitory capacity. In the first experiment we considered, Wasserman *et al.* (1974) observed a very distinct response to an E1 exposed in an E1 → no E2 association. Recall that these experimenters presented pigeons with a negative correlation between the illumination of a localized key and the presentation of food so that the delivery of food never occurred during the illumination of the key nor shortly after it. As a result of this experience, the pigeons actively withdrew from the side of the chamber in which the lighted key was presented. This withdrawal response cannot be viewed simply as the result of the inhibition of some other behaviour. Even if the pigeons had an initial tendency to approach the lighted key, the inhibition of such an approach response would leave the birds indifferent to the locus of the key rather than causing them to withdraw from it.

The capacity of E1 to elicit a particular response pattern after E1 → no E2 training is much more readily explained in terms of the second representational structure we considered. This model argued that exposure to an E1 → no E2 relationship forms an excitatory link between the E1 representation and another element which encodes something about the omission of E2. We can refer to this as a no-E2 representation. Presentation of E1 now excites this no-E2 element which in turn produces the observed behavioural response. Given this type of structure the question that now faces us concerns the nature of the information about the omission of E2 which is encoded by this representation. We can make a start on answering this question by looking at the type of behaviour controlled by E1.

In the Wasserman *et al.* experiment a lighted key paired with the omission of food elicited a withdrawal response, which suggests that the key had aquired aversive properties for the pigeons. Perhaps such aversive properties are a general feature of events associated with the omission of an attractive stimulus, such as food. In the previous section we discussed an experiment by Hyde (1976) demonstrating that exposure to a tone → food association made the tone an attractive stimulus to the animals. After exposure to this association, the rats performed a lever-press response to produce the tone just as they would respond for food itself. The same test could be used to investigate the motivational significance of the tone after exposure to a tone → no food relationship. Hyde (1976) also performed such an experiment. Initially he programmed a tone → no food relationship by arranging for food to be delivered on average once every 30 seconds except when a 3-second tone came on. Food was never delivered during either the tone or a 4-minute period after each tone presentation. In the second stage the rats were presented with a lever and each press produced the tone. The rate of lever-pressing in the group which had experienced the negative correlation between the tone and food was markedly lower than that of a group receiving random presentations of the tone and food. Clearly the tone, by being presented in a tone → no food association, had acquired aversive properties in complete contrast to the case where exposure to the tone → food relationship rendered the tone an attractive stimulus.

The obvious explanation of this effect is that the omission of an unexpected attractive stimulus, such as food for hungry animals, is itself an aversive event (Wagner, 1969b; Daly 1974), and we can assume that this event will activate some representation encoding the aversive nature of the omission of the expected attractive stimulus. By establishing a tone → no food association we ensure that the presentation of the tone is paired with the activation of this aversive representation, thus setting up a representation of the form 'the tone causes an aversive event'. In terms of our excitatory-link model, such a representation would be embodied by an excitatory link between the tone and the aversive event representation activated by the omission of food.

But what sort of information does this aversive representation encode? Does it, for instance, encode any information about the sensory-perceptual properties of the omitted E2, in this case food? In

Table 8. *Design of the Dickinson and Dearing (1979) experiment*

Groups	Stage 1	Stage 2	Test	Suppression ratio
E	clicker → food clicker + light → no food	tone + light → shock	tone	0.15
R	clicker → food light/food	tone + light → shock	tone	0.07
C	clicker → food	tone + light → shock	tone	0.01

the previous section we saw that under certain circumstances exposure to an E1 → E2 relationship could set up a representation which predominantly specified information about the motivational properties of E2. The strongest evidence for a purely motivational representation came from the Bakal *et al.* demonstration of transreinforcer blocking between a shock and a klaxon. This blocking effect suggested that the shock and the klaxon, although very different in terms of their sensory-perceptual properties, could be encoded in terms of a common representation which just specified the aversive nature of these events. Perhaps the omission of an attractive stimulus also activates the same motivational representation as the presentation of an explicit aversive stimulus as shock. An obvious way to answer this question is by looking for the appropriate transreinforcer blocking effect.

I have attempted to demonstrate such a transreinforcer blocking effect in rats (Dickinson & Dearing, 1979). The design of the experiment is illustrated in table 8. In the first stage Group E received a schedule designed to establish a light → no food association. These rats received a series of trials in which a clicker was paired with food intermixed with trials in which a compound of the clicker and the light was presented without food. This schedule ensured that the light was paired with the omission of food expected on the basis of the presence of the clicker. The control groups, Groups R and C, both received the same pairings of the clicker and food. In addition Group R also experienced the same number of presentations of the light as Group E, but in this case they were randomly related to the delivery of food. Group C were never exposed to the light during this stage. In the second stage all groups received a series of trials in which a compound of the light and a novel tone terminated with the delivery of a shock. Finally, the amount that the animals had learned about the tone → shock association in Stage 2 was measured by presenting the tone alone in the final test stage and seeing how much it suppressed lever-pressing for food.

If establishing a light → no food association caused the light to activate the same general representation as that excited by the shock, the occurrence of the shock in Stage 2 should have been relatively unsurprising for Group E. This means that Group E should have learned less about the tone → shock relationship during the trials in which the tone–light compound was paired with the shock. Table 8 shows that such a blocking effect did occur; on test the tone produced less suppression in Group E than in the control groups. The implication is that the omission of expected food and the presentation of shock excite a common representation. As these two events could not be more contrasted in terms of their sensory-perceptual properties, this common representation must encode information about the motivational significance of these events. This idea makes some rather counterintuitive claims about the nature of E1 → no E2 learning. Essentially, it argues that the animal learns that 'E1 causes an aversive event', but not that 'E1 causes the omission of a particular food'. Although animals are clearly sensitive to the omission of expected events and respond to them by generating an emotional or motivational state, the implication is that they have no way of representing the non-occurrence of a particular event. This is, of course, a very severe restriction on the cognitive powers of animals.

Unlike the inhibitory-link theory, the present model provides no explanation of why exposure to an E1 → no E2 association should endow E1 with inhibitory properties, and yet the ability of E1 to inhibit or reduce the responses elicited by another stimulus, paired with E2, represents one of its cardinal properties. We have already seen that an E1 paired with the omission of an attractive E2 acquires aversive properties. The question is whether we need to appeal to anything more than aversiveness to explain the inhibitory capacity. If E1 is inhibitory because it is aversive, then any other aversive stimulus should exert the same inhibitory effects. There is considerable evidence in fact that an aversive E1, established by pairing this stimulus with an explicit aversive E2, such as a shock, inhibits responses maintained by an attractive stimulus (Dickinson & Pearce, 1977; Dickinson & Dearing, 1979). In fact, we have already discussed examples of such an inhibitory influence extensively without drawing attention to its nature. This inhibitory effect is seen in a conditioned suppression experiment. In this procedure it is found that a stimulus paired with an aversive E2, such as a shock, suppresses lever-pressing for food and licking for water. This suppression is a clear example of a case in which an aversive stimulus inhibits behaviour maintained by an attractive one, the food or water. This inhibitory effect can be incorporated within our model by assuming that there is some antagonistic relationship between the motivational representations aroused by attractive and aversive events. It is a common observation that motivational and emotional states of contrasted affective value are incompatible. It is usually impossible to feel happy and elated while in a state of strong anxiety or fear. Given we accept that there is an intrinsic antagonism between motivational states of opposite affective polarity, the inhibitory properties of a stimulus, paired with the omission of an attractive E2, may arise directly from its capacity to arouse an aversive representation.

Our discussion so far has concentrated exclusively on the properties of E1s associated with the omission of attractive E2s. How, though, should we characterize learning about an E1 → no E2 relationship when E2 is an aversive event? If E1 activates an aversive representation when E2 is attractive, we might expect E1 to excite an attractive representation when E2 is aversive. Although we shall not discuss the relevant experiments here, what evidence we have suggests that an E1, paired with the omission of an aversive E2, acquires attractive properties. Just as an E1 associated with the omission of an attractive event will act as a punisher, so an E1 paired with the omission of shock will act as a reward (Rescorla, 1969b; Weisman & Litner, 1969). Thus once again we can assume that an E1 → no E2 relationship is encoded by an excitatory link between the E1 representation and another purely motivational representation encoding the fact that the omission of an aversive stimulus is an attractive event.

The conclusion we have come to about E1 → no E2 learning is far from being intuitively obvious. I have been arguing that the properties of E1 after such learning only require us to assume a representational structure of the form 'E1 causes an attractive (or aversive) event' and the implication is that, in contrast to certain cases of E1 → E2 learning, the animals do not encode anything about the specific properties of the omitted event. However, we must always bear in mind that this restriction might well reflect the limitations of our current techniques for interrogating the animal mind rather than a real restriction on their representational powers.

Action–E2 associations

In chapter 1 it was pointed out that event relationships differ in terms of the nature of the constituent events. We distinguished between a stimulus–E2 association, in which E1 is some event that is not under the animal's control, such as an environmental stimulus, and an action–E2 association, in which E1 is a component of the animal's own behaviour. Throughout this book we have been primarily concerned with learning about stimulus–E2 associations on the assumption that the conditions and mechanisms of learning are the same in the two cases (see page 36). In this chapter, however, we have concentrated on the form of the response changes produced by learning in the hope that these changes will give us a clue to the nature of the underlying representations. As the form of the behavioural changes observed during stimulus–E2 and action–E2 learning are so different, we must consider whether the representational structures we have developed so far will also account for the response changes produced by action–E2 learning.

Learning about a stimulus–E2 association is indexed by a change in the responses elicited by the stimulus in a Pavlovian or classical conditioning procedure. By contrast, learning about an action–E2 relationship is manifest by a change in the rate at which the animal performs the action itself in an instrumental or operant conditioning procedure (see p. 19). When E2 is an attractive event, such as food for a hungry rat, arranging an action → E2 relationship will lead to an increase in the rate at which the action is performed. So, for example, if we present a hungry rat with a lever-press → food association, the rat will come to perform lever-presses at a progressively faster rate.

Traditionally, American learning theorists, such as Hull and Guthrie, have always viewed the representation underlying this type of behavioural change as procedural in form. They assume that exposure to a lever-press → food association sets up a representation of the form 'when the lever is present, press the lever'. In fact they express this procedural representation in terms of an excitatory-link model so that the learning experience sets up a link between an internal representation of the contextual cues present at the time of training, such as the lever itself, and the mechanism which generates the lever-press response. This is, of course, the traditional stimulus–response, or S–R, theory which we have already briefly discussed. When such a representation has been formed, placing the animal in the presence of the appropriate contextual cues automatically triggers the response-generating mechanism via the excitatory link.

This theory has always met with resistance, and, when we think about it, S–R theory makes some very counterintuitive claims about the nature of the knowledge upon which instrumental action is based. The cardinal feature of an instrumental relationship is that it permits the animal to bring about a change in the environment by its own action. As a result, instrumental responses appear to be purposeful, being directed at the goal of achieving the environmental change. When observing a rat pressing a lever for food, we should explain this behaviour intuitively by saying that the rat wants food and knows that lever-pressing produces food. And yet it is the role of just this type of knowledge which S–R theory denies. The simple S–R structure contains no representation of the goal of the action, in this case food, and does not provide the animal with any mechanism for recording the outcome of its actions.

The obvious way to find out whether information about the goal of action is represented in the underlying associative structure is to look at an animal's integrative capacity. A large number of experiments, generally referred to as Pavlovian–instrumental transfer studies have attempted to do just this in one form or another (see Mackintosh, 1974; Rescorla & Solomon, 1967; Trapold & Overmier, 1972). However, I cannot possibly attempt an extensive review in this book, and I shall consider just one study, as yet unpublished, by C. D. Adams, because it represents a close parallel to the Holland and Straub (1979) experiment on stimulus → food learning which we have already discussed (see p. 76).

Initially Adams exposed hungry rats to a lever-press → food association. In this first stage rats were allowed to press a lever in an operant chamber 100 times and each press was followed by the delivery of a single sucrose pellet in a food magazine. In the second stage the lever was removed and the rats placed in the chamber for half-hour periods each day. On alternate days 50 sucrose pellets were delivered. In Group E-100 the rats were injected with lithium chloride following removal from the chambers on days on which they had received the sucrose pellets, but not on days without food delivery. The control group, Group C-100, received the lithium chloride injection following removal from the chamber on days on which no pellets were delivered. As a result of this training, Group E-100 was exposed to a sucrose → illness association and Group C-100 to a sucrose → no illness relationship. This stage of training continued until rats in Group E-100 refused to eat the sucrose pellets.

The third stage was designed to test whether these rats could integrate information about sucrose → illness association with that about the lever-press → sucrose relationship. Such integration could only occur if the associative structure set up by the lever-press → sucrose relationship contained a representation of the sucrose for this is the event in common between the two associations. If

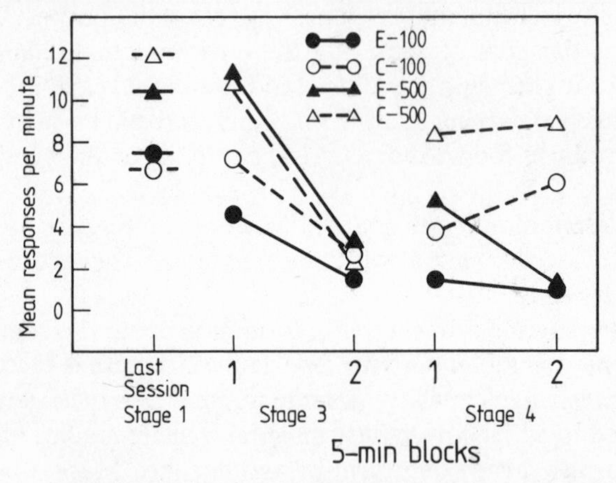

Fig. 23. The rats were initially trained to press a lever for sucrose pellets in Stage 1 during which they made either 100 (Groups E-100 and C-100) or 500 lever-presses (Groups E-500 and C-500). The figure illustrates the response rates of the various groups in the last session of Stage 1. In Stages 3 and 4 the rats were allowed to lever-press after the sucrose pellets had been paired with the induction of illness in Stage 2 for Groups E-100 and E-500 but not for Groups C-100 and C-500. The figure illustrates the response rates of the various groups when lever-pressing did not produce sucrose pellets in Stage 3 but did so in Stage 4.

the integration occurred, Group E-100 should have been less ready to press the lever when given the opportunity to do so than Group C-100. So the lever was returned in the third stage and the rate at which the rats pressed in the absence of sucrose pellets was measured.

Figure 23 shows the response rate of the two groups during the final session of training in Stage 1 before devaluation of the sucrose pellets and also during the test session in Stage 3. Although there was no difference between the response rates at the end of Stage 1, Group C-100 clearly responded more than Group E-100 in Stage 3. The rats appeared to be perfectly capable of integrating action → food and food → illness associations. This integrative capacity, however, cannot be encompassed by a simple procedural model of the S–R form and points to a role for a declarative type of representation for instrumental learning.

Once again, however, we cannot make a universal claim about the nature of associative representations. Earlier in this chapter we saw that extending the length of the chain of integration by using a higher-order conditioning procedure brought about failures to integrate separate, but relevant, stimulus–E2 associations. As yet the effect of comparable manipulations on the integrations involving action–E2 associations has not been looked at in detail. But Adams did study the effect of another variable which, on intuitive grounds, might be expected to result in the formation of a procedural representation. It is often observed that a transition in control occurs during the acquisition and practice of a coordinated system of instrumental acts. The novice car driver will start out by being told that pulling the gear lever in a certain direction engages top gear. This information may well be stored in a declarative form such as 'pulling back the gear lever engages top gear'. Thus when the engine revolutions (or the instructor) indicates that top gear is required, the novice can interrogate the declarative representation to derive the fact that a certain movement of the gear lever is required. The skilled driver, by contrast, does not appear to go through such a process; rather the appropriate action occurs whenever the right conditions are present. Thus for the skilled driver the knowledge about the conditions for gear changing are probably represented in the form 'when the engine revolutions are high, pull back gear lever', or in other words, by a procedural representation. Extended practice of an instrumental act seems to produce a transition of control from the declarative to the procedural form, thereby setting up a habit. A habit can be viewed as an action we perform automatically in a given situation without direct reference to the goal of that action.

Given such informal observations, Adams asked whether a similar transition would occur in the case where E1 is the act of pressing a lever. He ran two further groups of rats, Group E-500 and C-500, under exactly the same conditions as Group E-100 and C-100 except that these two groups received 500 rather than 100 lever-press–sucrose pairings in Stage 1. The question of interest is whether this extra training would render lever-pressing impervious to devaluation of the sucrose in Stage 2. As figure 23 shows, there was no difference between the rate at which Groups E-500 and C-500 responded during the test in Stage 3; the control of the lever-press response appeared to be autonomous of the value of the sucrose goal. In order to check that the sucrose pellets really had been devalued, Adams ran a fourth stage in which each lever-press again produced a sucrose pellet. In this stage responding was restored in Group C-500, but not in Group E-500, showing that the sucrose pellets were indeed unattractive to this latter group.

Extending the training from 100 to 500 action–sucrose pairings appeared to lead to a change in the nature of the representation controlling the instrumental behaviour. With low levels of training the representation could take part in integrative processes, a fact which points to a declarative form; however, increasing the amount of training rendered the response impervious to devaluation of the goal. Following our discussion of the representation set up by stimulus–E2 learning, we can offer two characterizations of the state of knowledge after extended training. First, it might be procedural in form. Alternatively, the transition might be between two declarative representations, one of which encodes the sensory-perceptual properties of the sucrose pellets and the other solely their motivational or affective value. As I have pointed out, we do not at present have any way of distinguishing between a procedural representation and a declarative structure which only encodes the affective value of E2 (see p. 94).

Having discussed the nature of the representations underlying action → E2 learning, we must also address the problem of the knowledge structures set up by action → no E2 relationships. We discussed stimulus → no E2 relationships extensively in the immediately preceding sections and came to the conclusion that such a relationship sets up an associative structure in which a representation of the stimulus is related to a purely motivational representation of the opposite affective value to that of E2 itself. The question is whether an analogous associative structure underlies the behavioural changes we see following learning about an action → no E2 relationship. One of the favourite procedures for studying such learning is an instrumental avoidance task in which E2 is an aversive stimulus, such as a shock. Let us start out by considering a typical example of such an avoidance schedule. Many avoidance experiments have used a procedure in which a rat is placed in a two-compartment chamber with an interconnecting door. This apparatus is called a shuttle box. Every so often a stimulus, say a noise, comes on and if the animal does nothing a shock is presented 10 seconds later. However, if during the initial 10 seconds of the noise the animal crosses from one compartment to the other, the noise goes off and the shock is omitted. After an intertrial interval the noise is again presented and the animal can avoid the shock by crossing from one compartment to another, and so by shuttling back and forth between the two compartments at the appropriate time the rat can avoid all the shocks. This procedure presents the animal with a typical E1 → no E2 association in which the shuttle response is E1 and the shock E2. Recall that the E1 → no E2 association requires that E1 is paired with the omission of an expected E2. The pairing of the noise and shock before the animal has acquired the shuttle response ensures that the shock is expected when the noise comes on. Thus a shuttle response during the noise is paired with the omission of the expected shock.

If we assume that the avoidance schedule sets up an associative structure similar to that seen following stimulus → no shock learning, we should expect the shuttling response to be based on a representation of the form 'shuttling causes an attractive event'. This means that performing the shuttle response should produce an attractive state in the animal and is clearly compatible with the fact that the animals emit this response. The problem, however, is to provide some independent evidence that performing the shuttling behaviour acquires attractive properties. We are arguing that the shuttling response acquires these properties because the avoidance schedule ensures that this behaviour acts as the E1 element of an E1 → no shock relationship. If this is true, we should be able to take any other potential E1, such as a light stimulus, and present it in exactly the same relationship to shock as the shuttle response. If the shuttle response acquires attractive properties on the avoidance schedule, then so should the light. Morris (1975) has attempted to find out whether this is so in an experiment with the design outlined in table 9. One group of rats, the master group, simply received avoidance training in the shuttle box in which presentations of the noise on trials when the rats failed to shuttle were paired with shock, whereas the compound of the noise and the shuttle response was paired with the omission of shock. During this first stage the remaining groups were placed in simple conditioning chambers in which an avoidance response was not possible. One rat in each of these remaining groups was paired or 'yoked' to a particular rat in the master group. Whenever the master animals failed to perform the shuttle response on a particular trial, the yoked animals in the other groups received a pairing of the noise and shock. However, on trials in which the master animal shuttled, and thus avoided the shock, the yoked animals received the same noise presentation as the master animals and no shock. In addition the rats in Groups L/L and L/N experienced a 10-second light which started at the time when the master

Table 9. *Design of the Morris (1975) experiment*

Groups	Stage 1	Stage 2
	shuttle box	
Master	noise → shock	
	noise + shuttle → no shock	
	Pavlovian chambers	*shuttle box*
L/L	noise → shock	noise → shock
	noise + light → no shock	noise + shuttle → light + no shock
L/N	noise → shock	noise → shock
	noise + light → no shock	noise + shuttle → no shock
N/L	noise → shock	noise → shock
	noise → no shock	noise + shuttle → light + no shock

animals performed the shuttle response. This ensured that the relationship between the onset of the light and the presentations of the noise and shock in Groups L/L and L/N was exactly the same as the relationship between the shuttle response and the noise and the shock in the master animals. As a result any properties acquired by the shuttle response for the master rats should have been acquired by the light for Groups L/L and L/N.

If the shuttle response acquired attractive properties during avoidance training, the light should acquire the same properties for Groups L/L and L/N. This means that the rats should have been prepared to perform a response which produced the light. To test whether this was so, Morris placed the yoked rats in a shuttle box in a second stage and trained them on an avoidance schedule (see table 9) in which Group L/L received the light immediately following each shuttle response, while Group L/N did not. If the light had become an attractive event, the rats in Group L/L should have performed the shuttle response more readily than animals in Group L/N. Figure 24 shows that this is exactly what happened; Group L/L shuttled on a higher percentage of trials than Group L/N. To ensure that the attractive properties of light depended upon experiencing the relationship between the light, noise, and shock, determined by the avoidance performance of the master animals in Stage 1, a third yoked group, Group N/L, did not receive the light in Stage 1 but experienced it following each shuttle response in Stage 2. Their performance was clearly inferior to that of Group L/L.

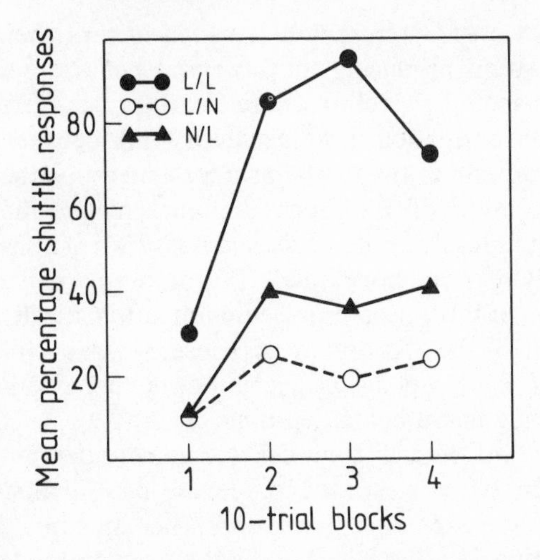

Fig. 24. The percentage of trials on which the rats made a shuttle response during the noise in Stage 2 and thus avoided the shock (see table 9). A shuttle response during the noise was followed by the presentation of a light in Groups L/L and N/L but not in Group L/N. The light had been previously exposed in a light → no shock relationship in Stage 1 for Groups L/L and L/N but not for Group N/L (see table 9). (After Morris, 1975.)

The sequence of events programmed by an avoidance schedule endowed a stimulus which bears the same correlation with the shock as the avoidance response with attractive properties. The implication is that the avoidance response itself acquired the same properties mediated by a representation of the form 'shuttling causes an attractive event'. Of course, if the response was paired with the omission of an attractive event, such as the delivery of food, rather than a shock, the action should acquire aversive properties.

Excitatory links and instrumental action

Although we may accept that instrumental learning reflects the operation of a procedural representation under some circumstances, Adams' experiment on the devaluation of rewards leaves us with little doubt that integration can occur and that some form of declarative representation is also required. In discussing stimulus → E2 learning we argued that most of the behavioural changes observed during such learning were compatible with a declarative representation in which an excitatory link was formed between the event representations of E1 and E2. When E1 is an environmental stimulus, presenting this stimulus activates its own event representation which in turn then excites the E2 representation producing the observed behavioural changes. By analogy, we should expect exposure to an action → E2 association to result in the formation of an excitatory link between the representation of this action or the mechanism for generating the action and the representation of E2. Such a model makes the plausible assumption that at least certain of the animal's own actions can be represented within its memory store. Furthermore, if we assume that the activation of this action representation automatically produces the performance of the appropriate response, we can give, what appears to be at first sight, a plausible account of the behavioural changes observed during instrumental learning.

To illustrate the operation of such a model, let us consider the simple case in which a hungry rat is exposed to a lever-press → food association. According to the excitatory-link theory, such an experience results in the formation of a link between the representations of the food and the action. Thus, when the food representation is activated, the response-generation mechanism will be excited and the lever-press occur. The question then is how does the food representation become aroused in the first place. The normal answer is to point to the fact that embedded within every action → E2 relationship is a stimulus → E2 association. Presenting food following each lever-press ensures that the delivery of food is paired with the contextual stimuli of the environment in which the response occurs. In the case of a typical instrumental lever-press conditioning experiment such stimuli would be the lever itself and the other features of the operant chamber. Thus an excitatory link would be formed between the representations of the contextual cues and that of the food, as well as between the food and the response-generation mechanism. Basically what this model maintains is that placing the animal in the operant chamber makes it think about the goal of the action, food, which in turn causes it to think about the response. This then results in the performance of the action.

This view of action → E2 learning has had a long history. It was originally proposed by Beritov and Pavlov, developed by other Russian theorists (see Gormezano & Tait, 1976), and subsequently taken up by certain American investigators (see Trapold & Overmier, 1972) who married it to a procedural or S–R theory of instrumental learning. In addition, the theory has made some striking predictions which have received empirical confirmation. Consider the following task. On each trial the rat is presented with two levers, one on the left and one on the right of the food magazine. On each trial pressing only one of the levers will produce food, and the side of the correct lever depends upon the nature of a stimulus present on that trial. Let us assume that when a clicker is on, the right-hand lever produces food, and when a tone is present the left-hand lever is correct. Thus the rat is required to learn two associations which we can characterize as follows: right-hand press (clicker) → food and left-hand press (tone) → food. The tone and the clicker are referred to as discriminative stimuli, which are stimuli that specify when a particular action–E2 relationship is in operation. According to the simple excitatory-link model of the underlying associative structures, outlined in figure 25A, the rats should be unable to solve this discrimination task. Both the clicker and tone representation should become connected to the food representation, which in turn will be connected to both the response representations. So presenting either the clicker or the tone will excite the food representation which will then excite both actions. Rats do in fact find such discrimination tasks very difficult, although not impossible. Perhaps the ability to finally solve such problems depends upon the formation of procedural representations of the form 'when the clicker is present, press the right-hand lever'.

As well as accounting for the difficulty of this task, the model does make some predictions about how to make this discrimination easier. For instance, the discrimination should be learned more rapidly if we followed correct responses on each lever by different types of food, say left-hand responses by food pellets and right-hand responses by sucrose solution. Now the two response representations can be linked to different food representations which can then be separately excited by the appropriate stimulus representations. This associative structure is illustrated in figure 25B. Trapold (1970) has shown that such a discrimination is learned more rapidly if responses on the two levers are followed by different types of food, rather than each response producing the same food.

This finding also allows for a further test of the model. Successful discriminative performance depends upon the animals learning about the clicker → food pellets and the tone → sucrose associations, for these are the first links of the associative structure. Trapold (1970) argued that if these associations were learned initially before any training with the levers, the rate at which the animals acquired the discrimination should be enhanced as one half of the associative structure would already be formed. In the first stage of a second experiment, Trapold exposed the rats to simple pairing of the clicker and sucrose and the tone and food in the absence of the levers. In the

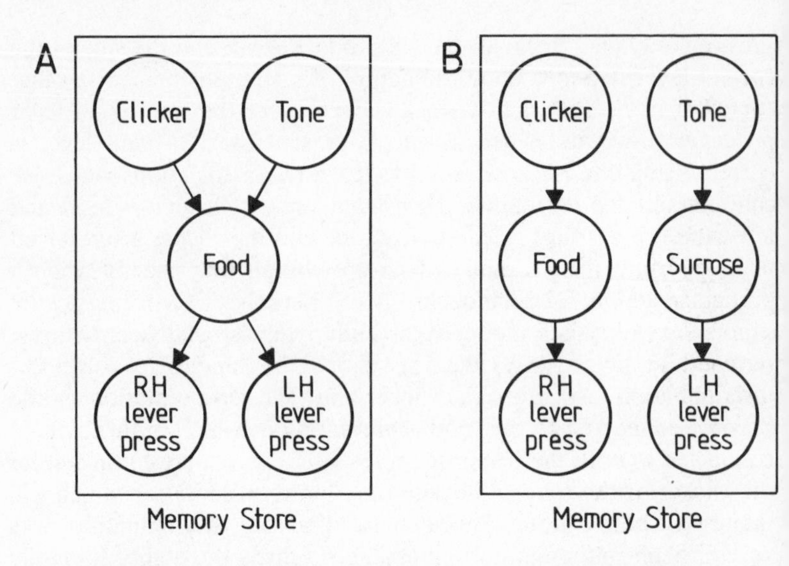

Fig. 25. An illustration of the possible associative structures set up by two discrimination procedures. Panel A corresponds to a procedure in which responding on the right-hand (RH) lever in the presence of a clicker, but not on the left-hand (LH) lever, is followed by food, and responding on the left-hand lever in the presence of a tone, but not the right-hand lever, is followed by food. Panel B corresponds to a similar procedure except that the left-hand lever-presses in the presence of the tone are followed by the delivery of a sucrose solution.

second stage the levers were introduced and one group of rats in the facilitation condition were trained on the right-hand press (clicker) → food pellets and the left-hand press (tone) → sucrose associations. Here the relationship between the stimuli and the rewards were congruent with those experienced during the first stage, and discrimination learning should have been facilitated. The second group in the interference condition was presented with right-hand press (clicker) → sucrose and left-hand press (tone) → food pellet relationships in the second stage. In this condition the associations between the stimuli and the rewards were incongruent in the two stages and discrimination learning should have been retarded. In line with this argument, figure 26 shows that acquisition of the discrimination was more rapid in the facilitation than in the interference condition.

In spite of these empirical successes, there are major problems with this excitatory-link model. The first concerns a hidden assumption that has crept into our exposition of this model. When animals are exposed to a stimulus → E2 relationship, the stimulus comes to elicit

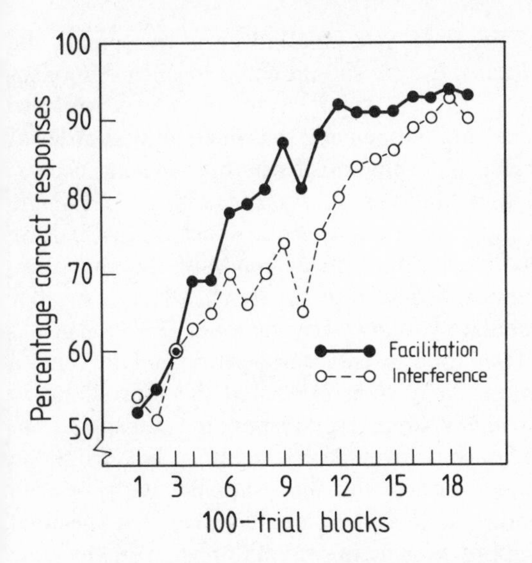

Fig. 26. The percentage of trials on which the rats responded on the correct lever in each session when they were trained on a discrimination procedure corresponding to the associative representations illustrated in Panel B of figure 25. Prior to the discrimination training rats in the facilitation condition were exposed to clicker → food and tone → sucrose associations which were congruent with the stimulus → E2 associations embedded within the discrimination procedure. By contrast, rats in the interference condition were pre-exposed to incongruent clicker → sucrose and tone → food associations. (After Trapold, 1970.)

a behavioural pattern appropriate to E2 (the principle of stimulus substitution). This we have taken as evidence that excitation can pass from the stimulus representation to the E2 representation. Thus we might argue analogously that exposure to an action → E2 relationship sets up a link which allows excitation to pass from the action to the E2 representation. But this is not the link illustrated in figure 25; here the excitation passes from the E2 to the action representation, and such a direction of activation is necessary if we are to explain why the action occurs. To maintain this type of model we have to assume that learning about an action → E2 relationship involves setting up the excitatory link from the E2 representation *to* the E1 representation when the response is E1. If such an associative structure is a general product of learning about E1 → E2 relationships, we should be able to observe manifestations of a link from the E2 to the E1 representation when E1 is a stimulus as well as an action. Consider the case in which E1 is a shock and E2 is a tone. If exposure to a shock → tone relationship, in which the onset of the shock preceded that of the tone, sets up a link from the tone *to* the shock representation then the tone should come to elicit responses appropriate to the shock, such as suppression. However, we have already seen that this does not happen and that the temporal order of the two events is a crucial determinant of whether learning occurs. Recall from chapter 2 that Mahoney & Ayres (1976) gave a single presentation of a 4-second tone and a 4-second shock and varied the temporal relationship between them. In one condition the tone came on first and was immediately followed by the shock, while in another the shock was first immediately followed by the tone. Thus one group of animals was exposed to a tone → shock association and the other a shock → tone association. They then presented the tone alone to measure the extent to which it suppressed responding. If exposure to the shock → tone relationship formed an excitatory link from the tone to the shock representation, the tone should have produced suppression in this condition. In fact figure 15 shows that the tone produced much less suppression in this condition than in the case where the animal is exposed to a tone → shock relationship. The obvious implication is that an E1 → E2 relationship forms a strong excitatory link only from the representation of E1 to that of E2, but not in the other direction. And yet it is exactly the reverse link which is required to explain the production of an instrumental action.

However, even if we accept that exposure to an action → E2 association sets up an excitatory link from the E2 representation to the action representation, the model still has problems. Consider the case of punishment. In a punishment procedure, the animal experiences an action → E2 relationship in which E2 is an aversive or noxious event, such as shock. According to the excitatory-link model, experience of this relationship should set up an excitatory link between the action and shock representations just as an action → food association forms a representational structure with an excitatory link between the action and food representations. Consequently, when the animal is placed in the environment in which this action has been punished, the shock representation should be aroused which in turn will activate the action representation producing the target behaviour. According to this model, animals should be just as ready to perform an action producing an aversive event, such as shock, as they are to respond for an attractive stimulus. Of course, this is not what happens; punishment procedures suppress responding. A number of theorists (e.g. Dinsmoor, 1978) have attempted to solve this type of problem by assuming that the response suppression seen during punishment does not really reflect the fact that the animals have learnt about the action → shock relationship, but rather is a by-product of some other process. However, all such accounts have, on closer inspection, turned out to be unsatisfactory as a general explanation (see Dunham, 1971; Mackintosh, 1974).

The core of the problem we are having in characterizing instrumental learning lies with the impoverished representational medium provided by the concept of an excitatory link. To illustrate this problem, let us try to put ourselves in the position of a hungry rat pressing a lever for food, and ask how we should account for the occurrence of this action, given, of course, that this behaviour has not become independent of the value of the food, or in other words a habit. A common-sense explanation of our action would probably start with the motivational statement or proposition that 'I want food'. Given this premise, an appropriate inference is a motivational command or instruction of the form 'perform a response that causes food'. This command taken together with the knowledge we have acquired about the lever-press → food association, a statement or proposition of the form 'lever-pressing causes food', would allow us to derive an action command 'perform lever-pressing' which can directly control our behaviour. Thus we can represent the chain of practical inference underlying our action in the following form:

'I want food' . . . . motivational proposition
therefore 'perform an action that causes food' . . . . motivational command
'lever-pressing causes food' . . . . knowledge proposition
therefore 'perform lever-pressing' . . . . action command.

An exactly analogous chain of inferences should be constructed for the case in which lever-pressing is punished by the delivery of shock:

'I don't like shock' . . . . motivational proposition
therefore 'do not perform an action that causes shock' . . . . motivational command
'lever-pressing causes shock' . . . . knowledge proposition
therefore 'do not perform lever-press' . . . . action command.

This common-sense view of instrumental performance may seem so obvious to us as to be entirely trivial. The important point, however, is that the chain of inferences involved in this type of explanation cannot be mediated by the excitatory-link model we have been considering. Remember that this model simply assumes that an E1 → E2 association is represented by an excitatory link between the event representations. This knowledge is translated into action by activating one of the representations, either directly by presenting the relevant event or via another excitatory link. Even if we set aside the lack of evidence for an excitatory link from the E2 to the action representation, this type of knowledge–action translation process cannot mediate the form of inference represented by the interaction of the motivational command and knowledge proposition to derive the action command for both reward and punishment. What we are essentially trying to capture by pointing to the role of inferential processes in instrumental performance is the idea that such action is in some sense rational and goal-directed. An instrumental action is not automatically either elicited or inhibited by the arousal of some other element of a chain of excitatory links; rather the control of action is determined by the interaction of the animal's knowledge about the consequences of its behaviour with the value it places on these consequences. It is this goal-directed character of true instrumental behaviour which the simple excitatory-link model cannot handle. However, we should not pretend that pointing to the role of inferential-like processes in the control of action in any way constitutes a theory of instrumental learning and performance. Although attempts have been made to formulate psychological theories along these lines (e.g. Goldman, 1970), I know of no account which is both capable of explaining the goal-directed nature of instrumental behaviour and open to empirical test. All we can do at present is to emphasize that declarative representations of an animal's knowledge about action–E2 relationships have to be in a form compatible with knowledge–action translation processes which capture the rational or inferential basis of true instrumental action.

Even if we have to abandon the excitatory-link model for instrumental learning, can we still maintain this model for the behavioural changes seen during stimulus–E2 learning, or in other words during classical or Pavlovian conditioning? On the face of it, the responses elicited by the stimulus exposed in a stimulus → E2 association do not appear to be goal-directed. Let us consider the first example we looked at in this chapter. Recall that Holland (1977) exposed hungry rats to pairings of a tone and food, and found that they started to move towards the magazine in which the food was delivered during the tone. The excitatory-link model claims that the rats show this approach behaviour because activation of the tone representation excites a representation of the delivery of food in the magazine, which in turn automatically elicits the magazine approach. According to this model, the behaviour is not controlled by any goal; it is just automatically elicited. And it is not obvious what goal the activity could have, for the food is delivered at the end of the tone whether or not the animal approaches during the tone.

However, perhaps there is a hidden goal for the approach response. The rat, of course, must approach the food magazine when the food is actually delivered in order to eat it, and it is possible that approaching prior to the delivery brings an advantage in that the animal is in some way prepared for the food delivery. What we are suggesting is that setting up the relationship between the tone and the food actually brings about another relationship between the approach response and the receipt of the food. This means that the behaviour we observe is not a direct consequence of the tone → food association, but rather reflects an instrumental approach → food relationship which is operative during the tone. Previously in this chapter we referred to this type of relationship as an approach (tone) → food association, and noted that the tone is then identified as a discriminative stimulus for the instrumental action of approach rather than as a Pavlovian conditioned stimulus. If the approach response was really controlled by this type of instrumental relationship, it would mean that exposure to this tone → food association must set up a representation which is compatible with the type of inferential process we have considered for instrumental performance. The implication is that perhaps all behavioural changes we observe during exposure to a stimulus → E2 association are really a function of hidden action (stimulus) → E2 relationship and essentially rational and goal-directed in nature.

How might we go about investigating this idea? One possibility is to put the tone–food and the approach–food relationships in opposition to each other. Consider the case in which each 10-second tone presentation terminates with the delivery of food unless the rats approach the magazine during the tone. If the rat approaches the magazine during the tone, the food is omitted at the end of the trial. This relationship is referred to as an omission schedule. The question we have to consider is whether the approach behaviour would be maintained by this schedule. On the initial trials the rat would not approach the magazine and so experience pairings of the tone and food on every trial. If the behaviour of the animal was controlled solely by the tone → food association, the rat would begin approaching the magazine when the tone came on. As a result food would be omitted at the end of the tone, thus establishing a tone → no food association, which in turn would lead to a decrement in the approach response. However, as soon as trials occurred once again without an approach response, the tone → food association should be reinstated, thus restoring the approach behaviour. By continuing this analysis, we can see that if the behaviour of the rat is solely determined by the tone–food relationships, the omission schedule should maintain the approach response, albeit at a lower level than that seen when the tone and food are consistently paired. In fact we can estimate how strong the approach response should be under the omission schedule by running a yoked-control group. The rats in the omission and yoked-control groups are paired up so that every time a tone is presented to the omission member of the pair, it is also presented to the yoked-control member. Similarly every time food is presented to omission member, so it is to the yoked-control member. This means that both rats of the pair receive exactly the same schedule of tone and food presentations. The only difference is that this schedule is entirely determined by the approach responses of the rat in the omission condition. If the approach response simply reflects learning about the tone–food association, the strength of this response should be identical in the omission and yoked-control rats as this relationship is the same for both animals.

By contrast, the idea that the approach response develops because of a hidden approach → food relationship makes exactly the opposite prediction. The omission condition ensures that the approach response is never followed by the delivery of food, so that approaching the magazine can be of no benefit to the animal. In fact the animal is exposed to an approach (tone) → no food association, and by exhibiting the response the rat actually prevents the delivery of food. In this sense the development of the approach responses cannot be regarded as goal-directed and should never occur in the first place.

Let us summarize these somewhat complex arguments. If the approach response is controlled solely by the tone–food relationships, the response should develop and be maintained at roughly the same level in the omission condition as in the yoked-control condition. By contrast, if the behaviour is entirely controlled by the approach–food relationships, the approach response should be absent in the omission condition. An intermediate result in which the approach response occurs in the omission condition, but at a lower level than in the yoked-control condition, would be expected if both relationships play a role.

Recently, Holland (1979) has compared the development of the approach response for rats trained on either the omission or yoked-control schedules. Figure 27 illustrates the results of this experiment. Surprisingly, rats in the omission condition acquired the approach response just as rapidly as animals in the yoked-control group.

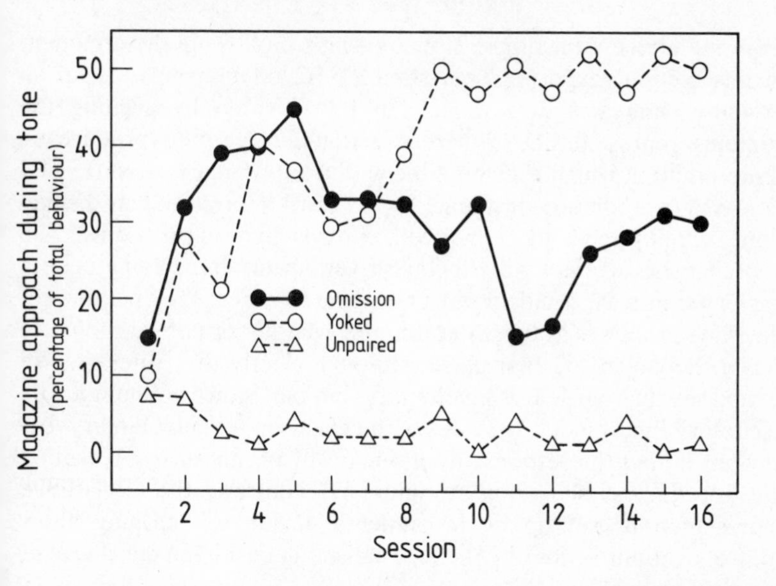

Fig. 27. The acquisition of magazine approach during a tone when the tone was paired with the delivery of food if the rat did not approach during the tone (omission condition), when the tone was paired with food irrespective of the behaviour of the rat during the tone (yoked condition), and when the tone and the delivery of food were unpaired (unpaired condition). The percentage of tone presentations paired with food and the temporal pattern of these pairings was the same in the omission and yoked conditions and determined by the approach behaviour of the rats in the omission condition. The strength of magazine approach is expressed by the percentage of all behaviours observed during the tone that consisted of approaching the magazine. (After Holland, 1979.)

Although the strength of the approach response in the omission groups dropped below that of the yoked-control group with extended training, it was still maintained well above the level of another control group, the unpaired group. This unpaired group received the same number of food and tone presentations as the rats in the omission condition but in the absence of any tone → food association. These results suggest that both the tone → food and approach (tone) → no food associations play a role in controlling the approach behaviour. However, from our point of view the really significant fact is that the response was maintained in spite of the omission schedule. It is as though the rat is to some extent unable to prevent the tone involuntarily eliciting magazine approach even when such behaviour is to its disadvantage.

The implication of Holland's (1979) experiment is that a stimulus–E2 association can produce behavioural changes in a way that differs from the effects of action–E2 relationships. Following the argument in this section, exposure to a stimulus–E2 relationship sets up an excitatory-link structure which can be activated by exciting the stimulus representation, whereas action–E2 learning produces a representation which is compatible with a knowledge–action translation system mediating or mimicking practical inference. This distinction is reinforced by our own introspective impressions. The apprehension we feel when being driven again after having been a passenger in a car accident does not seem to reflect the operation of the same processes as those determining whether or not we choose to get in the car in the first place, although clearly they interact. We should say that our fear is involuntary, but our choice voluntary. The excitatory-link model provides a mechanism for involuntary behaviour in that the response is automatically produced by presenting the appropriate activating stimulus. By contrast, the declarative representations set up by instrumental action–E2 learning allow control over our actions by the mechanisms determining the choice of our current goal.

Summary

This chapter has been concerned with the way in which animals represent information about the relationships between events. I started out by drawing a distinction between declarative and procedural representations. The declarative form corresponds to a description or statement of the relationship between the constituent events, whereas the procedural representation specifies the conditions under which the action or behaviour pattern produced by the association should be performed. The advantage of the declarative representation lies in the fact that it is easily interfaced with processes operating to integrate information about separate, but relevant, associations. Such integration does occur for both stimulus–E2 and action–E2 relationships. However, failures of integration are seen under certain circumstances and point to the possibility of an underlying procedural representation.

It was emphasized that any theory which attempts to delineate the specific nature of an associative representation must be accompanied by an account of how the knowledge incorporated in that representation is translated into the observed behaviour. The behavioural changes seen during stimulus → E2 learning can be described by the principle of stimulus substitution and are compatible with an excitatory-link model. According to this model, exposure to an E1 → E2 association sets up an excitatory link between the E1 and E2 representations, so that when the E1 representation is activated the E2 representation is also excited. Deviations from a strict principle of stimulus substitution can be explained by the fact that certain behaviours require the appropriate stimulus support from E1. In addition, we considered the possibility that the E2 representation may encode only certain of the properties of E2. Under certain circumstances, only information about the affective or motivational value of E2 appears to be encoded.

Exposure to an E1 → no E2 association endows E1 with motivational properties of the opposite affective polarity to that of E2. Thus, if E2 is aversive, E1 becomes attractive, and vice-versa. The inhibitory properties of E1 can then be explained by assuming that there is an antagonism between attractive and aversive motivational representations or states. As yet, there is no compelling evidence that E1 → no E2 learning sets up a representation which encodes information about the specific sensory and perceptual properties of E2.

Finally, we discussed the representation underlying action → E2 learning. The excitatory-link model faces two problems First it requires the presence of an excitatory link from the E2 to the E1 representation. However, evidence from backward conditioning experiments suggests that, at best, an E1 → E2 association sets up a very weak link of this kind. In addition, the excitatory link model has difficulty in explaining the response suppression seen under punishment. These considerations forced us to the view that stimulus–E2 and action–E2 associations set up qualitatively different types of representations, operating in conjunction with different knowledge–action translation processes. This view was supported by considering the effects of omission procedures. However, the fact that representations formed by stimulus–E2 and action–E2 associations are different does not imply that the learning mechanisms underlying the formation of these representations differ. In chapter 1 we saw that the conditions of learning appear to be the same when E1 is either an action or stimulus. This suggests the learning mechanisms are also similar in the two cases. We turn to a discussion of these mechanisms in the next chapter.

Before we finally leave the topic of knowledge representation, I should say a word about the possibility that animals can learn that two events are unrelated. In chapter 2 we discussed experiments by Baker and Mackintosh (1977) and Jackson *et al.* (1978) suggesting that animals could learn that E1 and E2 are unrelated when they are exposed to a zero correlation between these events. Although this learning is behaviourally silent, pre-exposure to such a random relationship retards learning about both E1 → E2 and E1 → no E2 associations when E1 is either a stimulus or response. Such learning provides profound difficulties for an excitatory-link model, for it is not obvious how to construct an associative representation of the absence of an event relationship in terms of excitatory links between event representations. This difficulty does not arise as acutely in the case of action–E2 learning for which we have already abandoned this type of representation. But what about stimulus–E2 learning? However, before we take this topic any further, we must discuss the mechanisms of learning. One account of the effects of exposure to a random E1–E2 relationship suggests that they are not mediated by an associative representation of the type we have been considering in this chapter, but rather by a change in the mechanisms governing the formation of the representation in the first place.

4 Mechanisms of learning

The time has now come to discuss the interface between the two aspects of learning which we have so far considered. We have seen that animals learn about event associations only under certain circumstances and that such learning can be viewed as the formation and strengthening of representations of these event relationships in the animal's memory store. In this chapter we shall discuss various theories about the nature of the mechanisms that ensure that these changes only occur when the animal is exposed to the appropriate conditions.

A prerequisite for our discussion of this problem is some general framework within which we can compare the different theories of learning that have emerged over the last decade or so, and I shall adopt the information-processing perspective popular in human cognitive psychology a few years ago (e.g. Atkinson & Shiffrin, 1971). Although most of the theories we shall consider were not initially developed and presented in terms of information-processing mechanisms, I hope that the use of this framework will not distort their central ideas. Figure 28 provides a schematic diagram of the

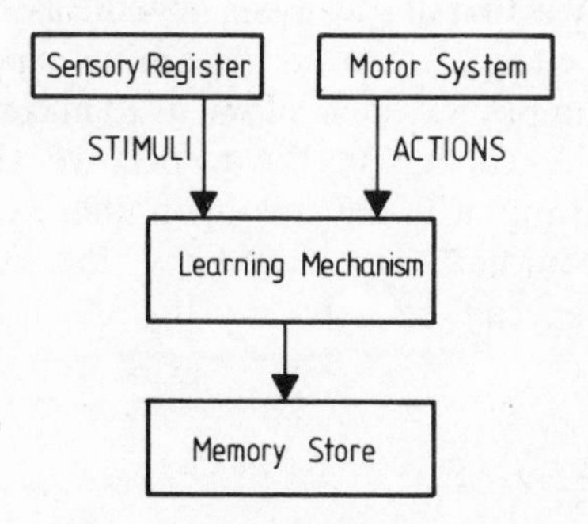

Fig. 28. A schematic illustration of the general learning model. Information about the occurrence of external stimuli provided by the sensory register and about the occurrence of actions provided by the motor system enters a learning mechanism. The extent to which information about the occurrence of these events is processed by the learning mechanism determines whether or not an associative representation corresponding to the relationship between events is set up or modified in the animal's memory store.

general model of learning which I shall adopt. Information about the occurrence of E1 and E2 enters some form of central learning mechanism or processor from a sensory register in the case of stimuli and from the motor system in the case of actions. If this information is processed in an appropriate manner by the learning mechanism, a change occurs in the representation of the relationship between the events in the memory store. Given this perspective, our task is that of characterizing the nature of the mechanism determining whether or not the events will be processed in a manner which leads to learning. Our guide in this task will be the conditions of learning outlined in chapter 1 for what we wish the mechanism to do is provide an explanation of why learning occurs only under the appropriate conditions.

Let us start out by considering the case of simple E1 → E2 learning. We saw in the first chapter that the conditions for such learning can be analysed in terms of the individual pairings of these events which the animal experiences. Rarely, however, does a single pairing appear to produce complete learning, and in numerous examples we have seen that the strength of the behavioural index of learning increases gradually across successive pairings. The progressive nature of these behavioural changes is assumed to reflect a gradual strengthening of the internal representation of the E1 → E2 association brought about by each pairing. The strength of this representation is usually referred to as the associative strength of E1 (for E2) and designated by the symbol V. Seen within these terms, the job of any learning theory is that of specifying how the associative strength of a particular E1 changes with each learning experience involving this event. An assumption, which allows us to make a start on this problem, is that the change in the associative strength on a particular trial or pairing will depend upon the extent to which E1 and E2 are conjointly processed by the learning mechanism. This simple idea can be expressed by the following equation:

$$dV = (\text{processing of E1})\ (\text{processing of E2}) \qquad 1$$

where dV is the change in the associative strength of E1 on that particular trial. Equation 1 is simply a formalization which captures the psychological idea that the amount an animal learns from a particular experience depends upon the degree to which information about E1 and E2 is conjointly processed. If either E1 or E2 fails to receive the appropriate processing, no change will occur in the associative strength of E1, or in other words in the underlying representation of the E1 → E2 relationship. Throughout our discussion of learning processes, we shall rely fairly heavily on formalizations developed from equation 1. It is important to realize, however, that such equations are only of use to the extent to which they make explicit in a concise form some underlying psychological idea about the nature of learning. It is this psychological theory which is of central importance, and not the particular formalization by which we choose to express it.

Given this general theoretical framework, the problem for any account of simple associative learning becomes that of specifying the factors that determine the processing of E1 and E2. One simple view would be that the processing of an event simply depends upon certain fixed properties of that event, such as its salience and intensity, and it is generally true that the more salient and intense a stimulus is, be it an E1 or E2, the more rapidly learning occurs. However, the salience and intensity of an event cannot be the only factors determining the amount of processing. In chapter 2 we saw that the ability of animals to track the correlation between two events depends critically upon the fact that simple pairings of E1 and E2 are not necessarily effective in bringing about learning. If their processing depended solely upon fixed properties of the events, each pairing should bring about the same amount of learning. We saw, however, that the amount an animal learns over a series of trials depends upon whether or not E2 is unpredicted or surprising. The role of surprise in learning is most clearly and simply illustrated by the Kamin blocking effect discussed in chapter 2. Recall that when a rat receives prior pairings of a tone and shock, the amount it learns about a light → shock association from pairings of a tone–light compound with shock is reduced; pretraining to the tone is said to block learning about the light → shock relationship. This simple blocking effect has become one of the main touchstones for evaluating competing ideas about learning mechanisms and I shall illustrate the operation of such mechanisms primarily in terms of this effect.

Why does E2 have to be surprising for learning to occur? There have been basically two types of answer to this question. The first assumes that an event must be surprising or unpredicted for it to be processed by the learning mechanism. The second position, on the other hand, argues that the processing of an event, rather than reflecting the extent to which it itself is predicted, depends upon whether it is a reliable predictor of other events. The bulk of this chapter will be spent in considering these two ideas.

Learning and event predictability

The general learning model, specified by equation 1, asserts that learning occurs as a result of a pairing of E1 and E2 to the extent that these two events are processed. The theories we shall consider in this section argue that the degree of processing depends not only upon the salience and intensity of E1 and E2, but also upon whether these events are unpredicted or surprising. In fact the most influential learning theory of the last decade, the Rescorla–Wagner theory, argues that most forms of selective learning, such as blocking, can be understood simply in terms of variations in the processing of E2 alone. We shall develop this idea first.

Processing of E2 – The Rescorla–Wagner theory

Although this exposition of the Rescorla–Wagner theory will not directly parallel their original presentation (Rescorla & Wagner, 1972), it will attempt to capture the fundamental ideas embodied in their theory. The theory can be viewed in terms of the model outlined in equation 1, namely that the increment in the associative strength of E1 depends upon the conjoint processing of E1 and E2. Rescorla and Wagner argue, however, that the processing of E2 does not simply reflect its salience and intensity, but also the extent to which it is unpredicted or surprising. The more surprising E2 is, the more it will be processed and hence the larger will be the increment in the associative strength of E1.

The problem we are now faced with is that of finding some way of specifying the degree to which E2 is surprising. Let us first consider a case involving a series of simple E1–E2 pairings. On any given pairing whether or not the animal expects E2, having been presented with E1, will depend upon whether the animal already knows about the relation, or in other words upon the current strength of the animal's internal representation of the E1 → E2 association. The strength of this representation, it will be recalled, is referred to as the associative strength of E1. The larger the value of this associative strength the less surprising will be the occurrence of E2. On the first trial or pairing, as the animal has not yet had the opportunity to learn about the relationship, the associative strength will be zero and E2 completely unexpected. Thus E2 will receive a certain amount of processing which in turn will lead to an increase in the associative strength of E1. The important question is how does this learning affect processing on the next trial. Recall that the central idea of the Rescorla–Wagner model is that the amount of processing received by

E2 decreases as the animal learns to expect it. Therefore we might anticipate that E2 processing on the second trial would be equal to the amount of processing when E2 is completely unpredicted minus the extent to which the animal expects E2, or in other words the associative strength acquired by E1 on the first trial. If we refer to the processing of E2 when it is unpredicted as λ and remember that the associative strength of E1 is designated by V, the processing of E2 on the second trial will be determined by the discrepancy $(\lambda - V)$. The more unexpected is the occurrence of E2, the smaller will be the associative strength of E1, and hence the larger will be the discrepancy $(\lambda - V)$. However, it is likely that other factors, such as the salience of E2, also play a role in determining its processing, and so it would be more realistic to argue that this processing is governed by a term such as $\beta(\lambda - V)$ where β represents the salience of E2. By reiterating this argument we can see that the processing of E2 on any particular trial will be given by the term $\beta(\lambda - V)$ where V is the associative strength of E1 on that trial.

Having determined the processing of E2, we can now focus on E1. A central feature of the Rescorla–Wagner model is the assumption that all variations in learning are due to changes in the effectiveness of E2 and that the contribution of E1 remains fixed. The implication is that the processing of E1 is the same on every trial and determined by some fixed property of this event, such as its salience. If we represent the salience of E1 by α, we are now in a position to substitute specific terms for the processing of E1 and E2 in equation 1 to give an expression which describes the change in the associative strength of E1 on a trial. Such a substitution yields:

$$dV = \alpha\beta(\lambda - V) \qquad 2$$

Equation 2 captures the central idea behind the Rescorla–Wagner model that the amount an animal learns from a pairing of E1 and E2 depends upon how surprising the animal finds the occurrence of E2.

In most of the experiments considered in this book, the amount learned on a particular trial is not usually measured directly; rather the course of learning over a series of trials is tracked or the amount learned after a number of pairings measured. Consequently, we should like the model to be able to specify the total amount learned after a certain number of trials. To track the course of learning over a series of trials, all we have to do is to add together the successive increments in the associative strength of E1 produced by each pairing. The associative strength of E1 after n pairings, V^n, will be given by the sum of the increment on the nth pairing, dV^n, and the total associative strength after n-1 pairings, V^{n-1}, so that:

$$V^n = V^{n-1} + dV^n \qquad 3$$

To illustrate the application of this model, let us consider a simple numerical example in which the amount of processing received by E2 when it is unpredicted (λ) and its salience (β) are both 1.0 and the salience of E1 (α) is 0.5. Before the first pairing the associative strength of E1 for E2, V^0, will of course be zero. Then, by applying equations 2 and 3, we can derive the associative strength after one pairing, V^1. This is given by:

$$\begin{aligned} V^1 &= V^0 + dV^1 \\ &= V^0 + \alpha\beta(\lambda - V^0) \\ &= 0 + 0.5 = 1.0 \times (1.0 - 0) \\ &= 0.5 \end{aligned} \qquad 4$$

Similarly after two pairings the associative strength, V^2, will be given by:

$$\begin{aligned} V^2 &= V^1 + dV^2 \\ &= V^1 + \alpha\beta(\lambda - V^1) \\ &= 0.5 + 0.5 \times 1.0 \times (1.0 - 0.5) \\ &= 0.75 \end{aligned} \qquad 5$$

By reiterating this process we can plot out the learning curve specifying the changes in the associative strength of E1 across successive pairings. This learning curve, illustrated in figure 29A, is negatively accelerated with the associative strength gradually approaching a maximum or asymptote equal to λ with successive pairings. The general shape of this curve roughly fits the way in which the strength or probability of a response changes during learning about a simple E1 $\rightarrow$ E2 association with the increments in response strength becoming progressively smaller with successive pairings.

We should not assume automatically, however, that the learning process can be observed directly by measuring changes of response strength; many other factors intervene to distort the way in which changes in associative strength are expressed in behaviour. Let us suppose that we are tracking changes in associative strength by measuring the strength of some response, and look at a number of factors which might prevent a direct translation of associative strength into response strength. A threshold might operate so that the associative strength has to reach a value of 0.5 before any change

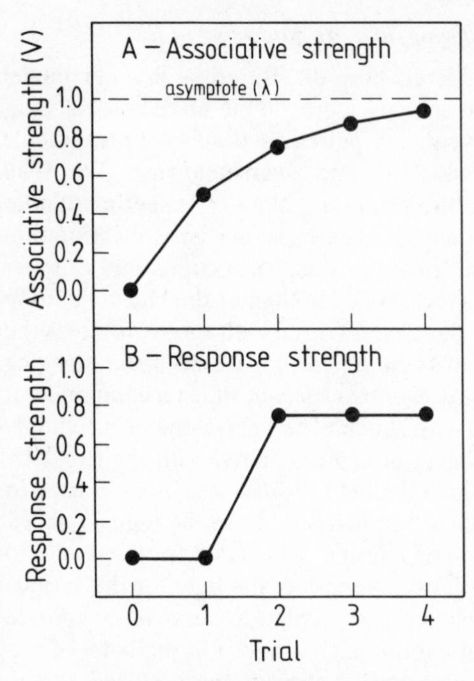

Fig. 29. Panel A illustrates the way in which the associative strength of E1 (for E2) changes across a series of pairings of E1 and E2 when these changes are governed by equation 2. Panel B shows the way in which changes in the associative strength illustrated in Panel A are translated into changes of response strength when threshold and ceiling factors operate.

in response strength occurs. In addition, it might turn out that, due to physical limitations, the animal cannot perform the response with a strength greater than 0.75. Even if we make the simple assumption that associative strength equals response strength once it is greater than the threshold, the operation of the threshold and ceiling factors will serve to distort the way in which changes in associative strength are manifest as changes in responding. Figure 29B illustrates the strength of the response produced on each pairing of E1 and E2 when learning follows the course shown in figure 29A. In practice we do not usually know the specific factors determining the relationship between associative and response strengths and as a result we have to test learning models by looking at predictions which do not depend critically upon the nature of this relationship. All we can assume is a monotonic function in which increases in associative strength never produce decreases in responding, and decreases in associative strength never produce increases in response strength.

Blocking, overshadowing, and the processing of E2

So far we have only considered how the Rescorla–Wagner model accounts for the fact that animals learn about an E1 → E2 relationship when these two events are paired, and now we must tackle the crucial problems of blocking and overshadowing. We shall illustrate the solution by considering an imaginary experiment consisting of three groups of animals. The basic design is illustrated in table 10. In this experiment a light, a tone, or a compound of these stimuli can act as E1 and a shock as E2. In Stage 1 the blocking group receives two trials in which the tone is paired with shock, whereas the other two groups receive no training. In Stage 2 both the blocking and overshadowing groups receive two trials in which a compound of the tone and light is paired with shock. The light-alone group simply experiences two trials in which the light is paired with the shock by itself. A final test trial in which the light is presented alone is used to measure how much the animal has learned about the light → shock association in Stage 2. Given this design, what we want the model to do is to determine the associative strength of the light for the various groups on test Trial 5 so that we can predict the relative magnitude the response elicited by the light in each group. On the basis of our discussion in chapter 2, we know that the response evoked by the light should be weaker in the blocking group than the overshadowing group; pretraining to the tone in Stage 1 will block learning about the light → shock association in Stage 2. In addition, the final response strength associated with the light in the overshadowing group should be less than that in the light-alone group, for the presence of the tone in Stage 2 should overshadow the light.

Table 10. *Design of a blocking and overshadowing experiment*

	Stage 1		Stage 2		Test
Groups	Trial 1	Trial 2	Trial 3	Trial 4	Trial 5
Blocking	T → sh	T → sh	LT → sh	LT → sh	L
Overshadowing			LT → sh	LT → sh	L
Light-alone			L → sh	L → sh	L

T: tone; L: light; sh: shock

Equations 2 and 3 describe the course of learning and to derive the final associative strength of the light, all we have to do is apply these equations separately for each group. However, we have to make some assumptions about the various parameters involved in these equations. For ease of presentation, let us assume that the parameters have the same values as in the previous example; the amount of processing received by the shock when it is totally unexpected (λ) is 1.0, the salience of the tone (α_T) and light (α_L) are both 0.5 and the salience of the shock (β) is 1.0. In addition we shall assume that the associative strengths of the light (V_L) and tone (V_T) are both zero before the start of training. Having made these assumptions, we shall start out by tracking the changes in the associative strength of the tone and light in the blocking group. On the two trials of Stage 1 the tone is paired with shock and the associative strength of the tone is given by equations 4 and 5 of our previous example. At the end of Stage 1 then, the associative strength of the tone after two trials, V_T^2, will be 0.75.

On the first trial of Stage 2, Trial 3, we meet a novel situation in that there are now two E1s present, the tone and the light. In order to handle this type of trial, we must go back to the basic principles of the Rescorla–Wagner model. Remember that the amount an animal learns from a pairing of E1 and E2 depends upon the extent to which E2 is processed. The strength of this processing in turn depends upon the difference between the degree to which E2 is processed when it is completely unexpected and the degree to which it is predicted by all the E1s present. When there is only a single E1 present, the degree to which E2 is predicted is given by the associative strength of that E1. When there is more than one E1, however, the degree to which E2 is predicted will be determined by the sum of the associative strengths of all the E1s present. As a result the processing of the shock when both the light and tone are present should be governed by the discrepancy ($\lambda - (V_T + V_L)$). We should therefore expect the increment in the associative strengths of the tone and light on Trial 3 to be:

$$dV_T^3 = \alpha_T\beta(\lambda - (V_T^2 + V_L^2)) \qquad 6$$

$$\text{and } dV_L^3 = \alpha_L\beta(\lambda - (V_T^2 + V_L^2)) \qquad 7$$

After Trial 2 the associative strength of the tone is 0.75 and that of the light is zero since the animals have never been previously trained to the light. By entering these values in equations 6 and 7 we get:

$$dV_T^3 = dV_L^3 = 0.5 \times 1.0 \times (1.0 - (0.75 + 0)) = 0.125$$

By adding these increments to the associative strength of the tone and light after Trial 2 we find that:

$$V_T^3 = V_T^2 + dV_T^3 = 0.75 + 0.125 = 0.875$$
$$\text{and } V_L^3 = V_L^2 + dV_L^3 = \quad 0 + 0.125 = 0.125$$

A similar application of equations 6 and 7 on Trial 4 gives:

$$dV_T^4 = dV_L^4 = 0.5 \times 1.0 \times (1.0 - (0.875 + 0.125)) = 0$$

So there is no further learning about the light → shock and tone → shock relationships on Trial 4, and at the end of the experiment the tone has an associative strength of 0.875 and the light of 0.125 in the blocking group. A graphic representation of the growth of the associative strengths is given in figure 30.

To track the growth of associative strength in the overshadowing group we simply have to follow the same strategy of applying

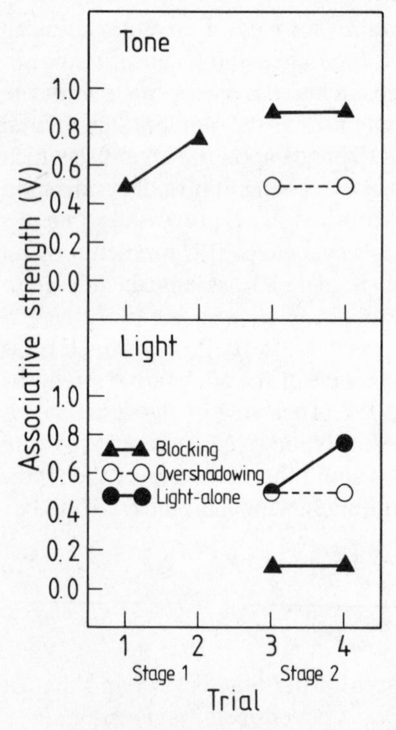

Fig. 30. The changes in the associative strengths of a tone (top panel) and a light (bottom panel) when the various groups are exposed to the blocking procedure illustrated in table 10. In Stage 1 the blocking group receives pairings of the tone and shock and in Stage 2 the blocking and overshadowing groups receive pairings of a tone–light compound and shock, while the light-alone group just receives pairing of the light and shock. The changes of associative strength are governed by equations 6 and 7.

equations 6 and 7, remembering of course that in this group the associative strength of the tone and light will both be zero on Trial 3 because this group received no training in Stage 1. Thus the increment in the associative strengths of the tone and light on Trial 3 will be:

$$dV_T^3 = dV_L^3 = 0.5 \times 1.0 \times (1.0 - (0 + 0)) = 0.5$$

After Trial 3 the associative strengths of the tone and light are both 0.5. A similar application on Trial 4 gives:

$$dV_T^4 = dV_L^4 = 0.5 \times 1.0 \times (1.0 - (0.5 + 0.5)) = 0$$

So no increments occur on Trial 4 and the final associative strengths of the light and the tone are 0.5. The basic blocking effect emerges because the final associative strength of the light on the test trial is only 0.125 in the blocking group, but 0.5 in the overshadowing group. This arises because training to the tone in Stage 1 increases its associative strength and thus decreases the size of the discrepancy $(\lambda - (V_T + V_L))$, a measure of how surprising E2 is, in Stage 2.

In the light-alone group only two trials are given with a single E1, the light, so that the associative strength of this stimulus is the same as that represented in equations 4 and 5. This means that on the test trial the associative strength for the light will be 0.75 in this group. This value is greater than the terminal associative strength of the light in the overshadowing group, which was only 0.5. This discrepancy represents the basic overshadowing effect.

Thus we can see that a formal statement of the Rescorla–Wagner model provides an explanation of the basic phenomena of overshadowing and blocking. In order to give a more general statement of the theory, which takes account of learning on compound trials on which there is more than one E1 present, we have to modify equation 2 along the lines suggested by equations 6 and 7. In such a modification the increment in the associative strength of a particular E1 is given by:

$$dV = \alpha\beta(\lambda - \Sigma V) \qquad 8$$

where ΣV is the sum of the associative strengths of all the E1s present on that trial for E2.

To ensure that we do not lose sight of the psychological import of the theory, it is worth restating the underlying principles of the model and the significance of each term in equation 8. The fundamental idea of the Rescorla–Wagner model is that the amount learned on a trial depends upon the degree to which the occurrence of E2 is predicted by all the E1s present on that trial. The better the occurrence of E2 is predicted, the less the animal learns. Blocking occurs because the pretraining to one E1 during Stage 1 ensures that the occurrence of E2 during Stage 2 is predicted so that the animal learns little about the second, added E1. The discrepancy $(\lambda - \Sigma V)$ represents a general measure of the extent to which E2 is surprising. In terms of our processing model, λ is a measure of the processing of E2 when it is totally unpredicted. Rescorla and Wagner (1972), however, initially did not present their theory in terms of such a model, and figures 29 and 30 point to a more neutral interpretation of this parameter. Figure 30 shows that the sum of the associative strengths of all E1s approaches the value of λ across successive trials and for this reason λ can be referred to as the asymptote of the associative strength. Because the final strength of a response elicited by an E1, and by implication its associative strength, often increases with the magnitude of E2, Rescorla and Wagner assume that the value of λ increases with the magnitude of E2. This assumption accords fairly well with our processing model, for it seems reasonable that the amount of processing received by an unpredicted E2 should increase with its magnitude. Finally the processing of E2 is also determined by its salience (β). As the rate at which the associative strength approaches the asymptote is determined by the value of β, this parameter can also be referred to as a learning-rate parameter.

The contribution of the processing of E1 to learning is assumed to be fixed and simply determined by its salience (α). When only a single E1 is present, the value of α will also determine the rate at which its associative strength approaches the asymptote and so can also be regarded as a learning-rate parameter. In compound training where there are two E1s, say X and Y, the amount learned about each E1 → E2 association will be determined by the processing of both X and Y. Equation 8 predicts that the final associative strength of X will be equal to $\lambda\alpha_X/(\alpha_X + \alpha_Y)$. When the salience of X and Y are equal, we should expect them to acquire equal associative strengths, which will be half that acquired when either is trained alone. As the salience of Y increases relative to X, Y should increasingly overshadow X, while X will lose its capacity to overshadow Y. These predictions are roughly in accord with the results of the overshadowing studies discussed in chapter 2.

Processing of E1–The Wagner theory

The Rescorla–Wagner model concentrates exclusively on the way in which variations in the processing of E2 affect learning; the processing of E1 is assumed to be fixed throughout learning, and yet we have already seen that changes in learning can be brought about simply by exposing the animals to E1. Two experiments by Baker and Mackintosh, discussed in chapter 2, showed that exposing rats to presentations of E1 retarded subsequent learning about both E1 → E2 and E1 → no E2 associations. In these experiments one group of rats, the tone-alone group, received a series of presentations of a tone by itself during the first stage. Another control group was placed in the experimental chambers for the same length of time without being exposed to the tone. In the second stage both groups of rats were required to learn about a tone → water association in one experiment and a tone → no water association in the other. In both cases pre-exposure to the tone retarded subsequent learning. It is difficult to see how this so-called latent-inhibition effect could be due to variations in the processing of E2 for the effect arises from changes which occur when E1 is presented alone. Rather, exposure to E1 appears to change its own processing in a way that reduces its subsequent associability with E2.

Wagner (1978) has recently extended the thinking behind the Rescorla–Wagner model of variations in the processing of E2 to encompass the idea that the effectiveness of E1 may also change during the course of learning. He argued that just as the processing of E2 depends upon whether or not it is surprising, so does that of E1. To the extent that the occurrence of E1 is predicted by other events and stimuli in the animal's environment, E1 will fail to be processed by the learning mechanism. The basic latent inhibition effect follows directly from this idea. During the pre-exposure phase the animal will experience pairings of the contextual cues and E1, and as a result the animal will learn about this context → E1 association. Thus, when the animals receive subsequent pairings of E1 and E2, the occurrence of E1 will not be surprising in this context and hence will receive less processing by the learning mechanism.

Once we have appreciated the possible symmetry in the mechanisms governing the processing of E1 and E2, the Rescorla–Wagner equation can be expanded to incorporate this symmetry. The processing of E1 should be affected not only by the salience of this event (α) but also by the extent to which it is predicted by the contextual cues, and thus we must incorporate a term in equation 8 to represent the role of this factor. Such a term will have the form $(\ell - \Sigma v)$ where ℓ is the extent to which a totally unexpected E1 will be processed and Σv is a measure of the degree to which all the contextual cues present predict the occurrence of E1, or in other words the combined associative strengths of all the stimuli present for E1. The inclusion of this term in the Rescorla–Wagner equation gives an expression of the following form for determining an increment in the associative strength of E1 for E2:

$$dV = \alpha(\ell - \Sigma v) \times \beta(\lambda - \Sigma V) \qquad 9$$

Although equation 9 may look somewhat complicated, all it essentially does is to capture in a formal manner the basic idea that the amount the animal learns about an E1 → E2 association depends upon the extent to which both E1 and E2 are unpredicted or surprising.

Capacity of the learning mechanism

Wagner (1978) has also extended the concept of a learning mechanism in one further aspect by suggesting that this mechanism has limited processing capacity, an idea which has been popular in discussions of human information processing. We have already considered the basic idea that the degree to which information about an event is processed depends upon whether or not it is surprising. The more surprising it is, the more likely it is to undergo the necessary processing for learning. However, if the learning mechanism has a limited capacity, presenting a sequence of surprising events may well interfere with learning by exceeding the processing capacity of the mechanism.

Wagner based this idea upon a rabbit eyelid conditioning experiment (Wagner, Rudy & Whitlow, 1973). In this experiment the rabbits were presented with a series of E1–E2 pairings and the rate of learning measured. Each pairing was followed by another event which was surprising for one group and predicted for the second. If the learning mechanism has a limited processing capacity, we might expect this post-trial event to occupy some of this capacity at the expense of E1 and E2 and thus interfere with learning about the E1 → E2 association on the immediately preceding trial. However, this interference should be observed only when the post-trial event is surprising, for only then will it take up significant processing capacity according to Wagner's theory. In the first stage Wagner and his colleagues established either a tone or a vibrotactile stimulus as a predictor of an eyeshock by exposing the animals to pairings of the relevant stimulus with the eyeshock. This predictive stimulus will be referred to as S+. The other stimulus, referred to as S−, was not established as a predictor and was simply presented alone in the absence of the shock. In the second stage they went on to measure the rate of learning about a light → shock association by presenting a series of trials in which a 1-second light was paired with the eyeshock. A post-trial event was presented by delivering a second eyeshock 11 seconds after each trial. For one group, Group S+/sh, the occurrence of this post-trial shock was rendered unsurprising by preceding its presentation by a 1-second presentation of S+. As soon as S+ occurred, the animal would expect the post-trial shock so that its occurrence should have been fully predicted. By contrast, for the second group, Group S−/sh, the occurrence of the post-trial shock was surprising as its presentation was preceded by S− which had never been previously associated with the shock.

Thus all the rabbits were given pairings of the light and eyeshock. The rapidity with which they learnt about the light → shock relationship should have depended upon the extent to which information about these events commanded the processing capacity of the learning mechanism. If the capacity of the mechanism is limited and the processing persists after the end of a trial, the presentation of another event, such as a post-trial shock, shortly after each light–shock pairing might disrupt learning. The processing of the post-trial shock should take up some of the capacity of the learning mechanism at the expense of the processing of the light–shock pairing on the immediately preceding trial. Thus the delivery of a post-trial shock may disrupt learning about the light → shock relationship. However, such disruption will depend upon whether or not the post-trial shock is surprising, for only if it is unpredicted will it command significant capacity in the learning mechanism. As a result we should expect to see little disruption of learning about the light → shock association if the post-trial shock is expected, as is the case where the shock is preceded by a previously established signal, S+. In line with this idea, figure 31 shows that the development of a conditioned eyeblink to the light was slower in Group S−/sh, for which the post-trial shock was not preceded by a previously established signal, than in the group in which such a signal was given, Group S+/sh.

This elegant experiment provides good evidence that the occurrence of a surprising event shortly after a learning trial interferes with learning on that trial, a finding clearly compatible with a learning mechanism of limited capacity. Furthermore the very sensitivity of learning to the unpredicted nature of post-trial events gives independent support to the idea that it is this feature of these events which is critical in determining their processing. Two further groups run by Wagner *et al.* reinforce this conclusion. Both groups received exactly the same sequence of light–shock pairings. For one group, Group

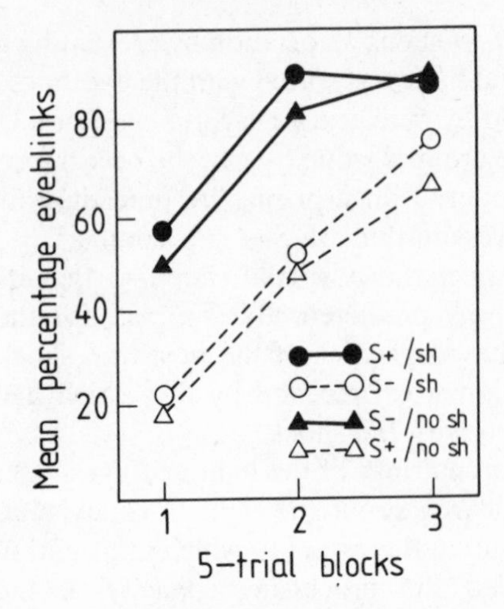

Fig. 31. The percentage of trials on which a light paired with an eyeshock elicited an eyeblink in rabbits. In Group S+/sh a second predicted post-trial shock was delivered 11 seconds after each light–shock pairing, whereas in Group S−/sh the post-trial shock was unpredicted. In the remaining groups the post-trial event consisted of either the surprising (Group S+/no sh) or expected omission (Group S−/no sh) of an eyeshock. (After Wagner *et al.*, 1973.)

S+/no sh, each pairing was followed 10 seconds later by the presentation of S+, whereas the second group received S− after each trial. However, no post-trial shock was presented. Thus Group S+/no sh received a surprising post-trial event in that the presentation of S+ should lead the rabbit to expect a shock, which in fact did not occur. The absence of a post-trial shock in Group S−/no sh, on the other hand, was fully predicted by the presentation of S−. Figure 31 shows that the unexpected omission of a post-trial shock in Group S+/no sh produced just as much interference with learning about the light → shock relationship as did the surprising presentation of a post-trial shock in Group S−/sh. It appears that the interference effect does not depend upon the physical properties of the post-trial event, but simply upon whether or not they are unexpected.

Finally this technique allows us to address another interesting question about the learning process. It is very unlikely that the processing of E1 and E2, involved in setting up a representation of the E1 → E2 association, is instantaneous, and we might well expect this processing to be extended in time after the end of the learning trial. Wagner *et al.* argued that such a temporal gradient of processing should be revealed by varying the time interval between the end of the learning trial and the delivery of a surprising post-trial event. We have already seen that if a surprising post-trial event occurs a few seconds after each light–shock pairing, the rate at which the rabbit learns is attenuated. If we increase this interval, however, until the processing of the light and shock on the preceding trial is complete, the delivery of a surprising post-trial event should be without effect. Wagner *et al.* varied the time between presenting each light–shock pairing and the delivery of a surprising post-trial shock between 4 and 301 seconds in different groups of rabbits. As figure 32 shows, the longer this interval, the more the rabbits learned about the light → shock relationship.

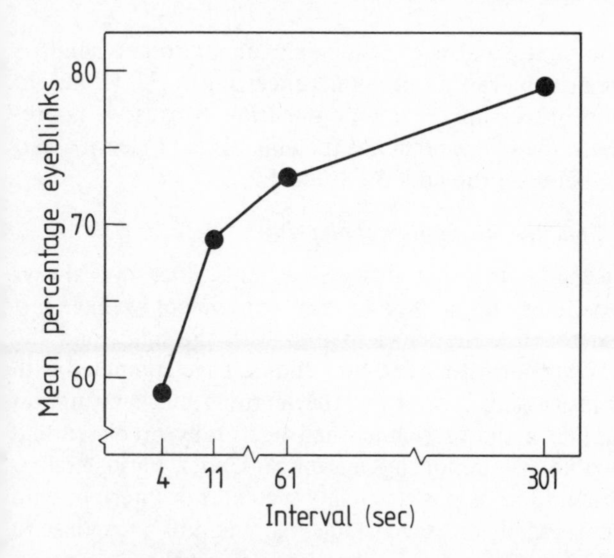

Fig. 32. The percentage of trials on which a light elicited an eyeblink over a series of pairings of the light and eyeshock. For all rabbits each pairing of the light and eyeshock was followed by a surprising post-trial event after an interval that differed for the various groups. (After Wagner *et al.*, 1973.)

In conclusion then, Wagner argues that selective learning depends upon variations in the processing of both E1 and E2. The more that an event, be it an E1 or E2, is predicted by others cues or events, the less likely the animal is to process an event and learn about a relationship involving it. In the latent inhibition, or E1 pre-exposure, procedure, subsequent learning is retarded because E1 becomes

Table 11. *Design of the Dickinson, Hall and Mackintosh (1976) experiment*

Groups	Stage 1 (12 trials)	Stage 2 (8 trials)	Test
0-0	T → sh	TL → sh	L
0-sh	T → sh	TL → sh-8-sh	L
sh-sh	T → sh-8-sh	TL → sh-8-sh	L
sh-0	T → sh-8-sh	TL → sh	L

T: auditory stimulus; L: light; sh: shock; sh-8-sh: shock followed by a second post-trial shock 8 seconds later.

predicted by the contextual cues. Blocking, on the other hand, is determined primarily by variations in the processing of E2. E2 fails to receive adequate processing on compound trials because it is predicted by the pretrained E1, and hence the animals fail to learn about the relationship between the added E1 and E2.

Surprise and the attenuation of blocking

The presentation of surprising post-trial events does not always interfere with learning. In the Wagner *et al.* experiment we have just considered both the E2, the shock paired with the light, and the post-trial shock were surprising and thus should have competed with each other for processing capacity in the learning mechanism. But what would happen if the first shock had been fully predicted and therefore incapable of supporting learning? Under these circumstances the presentation of post-trial shock could not interfere with learning and may actually enhance the development of a response to that target E1 by allowing the animal to learn about the E1 → post-trial shock relationship. Evidence for such enhancement comes from a study by Dickinson, Hall and Mackintosh (1976) who, like Wagner *et al.* (1973), investigated the effect of giving surprising post-trial events. However, in this study a blocking, rather than simply conditioning, procedure was used. The basic design is illustrated in table 11. The experiment employed four groups of rats and used a conditioned suppression procedure with a shock as E2. In Stage 1 all the rats received 12 trials with an auditory stimulus as E1 followed by 8 trials with an auditory–light compound stimulus as E1 in Stage 2. Finally the amount that the animals had learned about the light → shock relationship was measured in a test stage in which the light was presented alone to see how much it suppressed responding for food.

For Group 0-0 a standard blocking procedure was used in which each trial or stimulus presentation terminated with a shock throughout both stages. As a result, we should expect pretraining to the auditory stimulus in Stage 1 to block learning about the light → shock relationship in Stage 2. As figure 33 shows, this is just what happened; the animals in Group 0-0 had a suppression ratio just below 0.5, indicating that the light failed to suppress responding. Group 0-sh received identical training to that experienced by Group 0-0 except that 8 seconds following each compound stimulus–shock pairing in Stage 2 a second, unpredicted post-trial shock was delivered. Thus the auditory–light stimulus compound was paired with

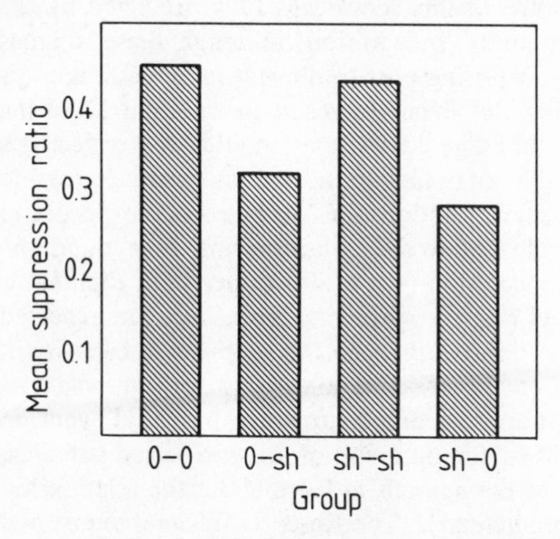

Fig. 33. The extent to which the presentation of a light suppressed lever-pressing for food on test in various groups of rats exposed to the blocking procedure outlined in table 11. Group 0–0 did not receive a post-trial shock during either stage, whereas Group sh-sh experienced post-trial shocks during both stages. Group 0-sh received post-trial shock only during Stage 2 and group sh-0 only during Stage 1. (After Dickinson *et al.*, 1976.)

two shocks with the post-trial shock being unpredicted by the auditory stimulus. In this case the post-trial shock cannot interfere with learning about the association between the light and the first shock because the blocking procedure ensures that no such learning occurs. In fact, to the extent that the animals learn about the relationship between the light and the post-trial shock, we should expect Group 0-sh to show some suppression to the light on test.

Figure 33 shows that Group 0-sh did in fact show more suppression to the light than Group 0-0. Of course, if the auditory stimulus was associated with both shocks during Stage 1, it should be capable of blocking learning about the relationship between the light and both the immediate and post-trial shock. Group sh-sh received a post-trial shock 8 seconds after the first shock on each trial of both stages, and in this group the light produced little suppression on test.

The Rescorla–Wagner theory appears to be able to explain the effects of a surprising post-trial shock in both simple conditioning and blocking procedures. However, let us now consider the final group in table 11, Group sh-0. These animals received the post-trial shock in Stage 1 but not in Stage 2. Once again the rats should have learned little about the association between the light and the first shock in Stage 2 as the occurrence of this shock was fully predicted by the pretrained auditory stimulus. In addition, although these animals were presented with a surprising post-trial event in Stage 2, there is no reason for expecting this event to result in the light acquiring suppressive capacities. In Stage 2 these rats should have expected a post-trial shock which did not in fact occur, and thus were exposed to a light → no post-trial shock relationship. We know that exposure to such a relationship should lead to the light acquiring, if anything, the capacity to alleviate suppression (see p. 18) rather than eliciting it. And yet figure 33 shows that the surprising omission of an expected post-trial shock in Group sh-0 enhanced the suppressive capacity of the light just as much on the presentation of a surprising post-trial shock in Group 0-sh. It appears that a surprising post-trial event, in this case brought about by the omission of an unexpected stimulus, can restore the ability of the animals to learn about the relationship between an E1 and a predicted E2. The Rescorla–Wagner theory and Wagner's extension of this model provides no ready explanation of this attenuation of blocking by post-trial surprise.

There is no doubt that the Rescorla–Wagner theory, especially in the form elaborated by Wagner, provides a powerful explanation of many of the cardinal features of simple associative learning, and over the last decade it has become the major theory in the area. However, certain aspects of the role of surprise in learning cannot be encompassed by this theory and should encourage us to look for alternative accounts. Before doing so, it is probably time to restate the overall purpose of all this analysis. By now it may appear that the whole topic has become enmeshed within a web of intricate procedural variations on a single experimental theme, blocking, of little general import or relevance. It should be remembered, however, that failures of learning, such as are seen in the blocking experiment, underlie the important capacity of animals, and probably ourselves, to detect the overall correlation between events. It is this ability which lies at the heart of the problem of tracking causal relationships.

Learning and predictive power

But why should animals want to track causal relationships? In chapter 1, I argued that such learning endows the animal with the capacity to predict, and thus anticipate, the occurrence (or non-occurrence) of important events. In the case of instrumental learning when E1 is an action, such knowledge allows the animal to gain control over its environment and through its own behaviour regulate the occurrence of important E2s. On intuitive grounds, this functional perspective suggests that the importance of an E1, and hence the degree of processing it receives, should be related to its power as a predictor of E2. But we have already seen that, in complete contrast to this general expectation, Wagner argues that the processing of E1 depends, not upon its predictive power, but upon the extent to which it itself is predicted by other events. The major alternative theories, however, describe the learning mechanism in terms which are more in accord with the apparent function of learning by assuming that the processing of an event depends upon its power as a predictor of other events.

The core of this idea can be illustrated by considering the latent inhibition phenomenon once again. Recall that during a latent inhibition experiment the animal is simply pre-exposed to a number of presentations of E1 alone before experiencing a relationship between E1 and E2. Such pre-exposure is found to retard learning about the E1–E2 association. A theory which maintains that the processing of E1 depends upon its predictive power would argue that during the pre-exposure phase the animal learns that E1 predicts nothing of significance. As a result its ability to command processing capacity in the learning mechanism is severely reduced, thus retarding any subsequent associative learning involving E1.

This general account has some merits. If we could ensure that E1 predicted an event of significance during the pre-exposure stage, it should not lose its capacity to be processed. One of my own experiments (Dickinson, 1976) suggests that this might be so. The study used a two-stage design with rats lever-pressing for food. In the second stage all three groups received pairings of a tone and shock, and the rate at which suppression developed to the tone was measured. The three groups varied in the conditions they experienced during the first stage. The paired group received a series of tone presentations during which food was presented freely to the animals at a fairly high rate, and so for this group the tone was presented in a predictive relationship to another event of significance, the delivery of food. As a result the tone should have retained the capacity to be processed when presented in the tone → shock association in the second stage. By contrast, the random group received the same number of tone and free-food presentations as the paired group in Stage 1, but in a random relationship to each other. Here the tone is not a predictor of any event of significance and so should lose its ability to command processing capacity in the learning mechanism. Thus these animals would be expected to learn less readily than the paired group about the tone → shock association in the second stage. Finally, a food-alone control group just received the free-food presentations in the first stage but no exposure to the tone. The rate of learning during Stage 2 in this group should provide a measure of the processing of the tone in the absence of any pre-exposure.

Figure 34 shows the development of suppression during Stage 2 when the tone was paired with the shock for all animals. The food-alone group learned very rapidly about the tone → shock relationship and exhibited considerable suppression on the second trial after only one tone–shock pairing. By contrast, the random group appeared to learn nothing on the first trial and showed minimal suppression on Trial 2. This difference represents the basic latent inhibition effect. The important results are those for the paired group. Although this group did not show as much suppression as the food-alone animals on Trial 2, they clearly learned something on the first trial in that their responding was more suppressed on the second trial than that of the random group. Pairing the tone with food during pre-exposure in Stage 1 attenuated the latent inhibition effect and by implication maintained the processing of E1. Ensuring that E1 is a good predictor of some other significant event, in this case food, appears to reduce the decline in the processing of E1 which would otherwise have occurred.

Limited-capacity processing theories

There have been a number of attempts to formulate accounts of why nonpredictive stimuli lose their associability. In the discussion of Wagner's theory I developed the common idea that the learning

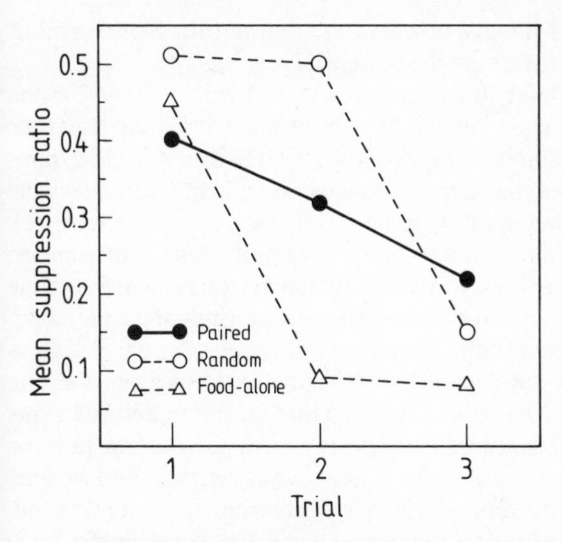

Fig. 34. The suppression of lever-pressing for food by rats produced by a tone during a series of pairings of the tone and shock. Prior to this training the paired group had received pairings of the tone and food, the random group uncorrelated presentations of the tone and food, and the food-alone group no pre-exposure to the tone. (After Dickinson, 1976.)

mechanism has a limited capacity. According to Wagner, access by information about an event to this mechanism for processing depends upon whether the event itself is predicted or not. A number of other theorists, while assuming that the capacity of the processing mechanism is limited, have argued just the opposite; namely that access depends upon the extent to which the event predicts others of significance. Moreover, given that the capacity of the system is limited, the presence of a predictive event will have consequences for the processing of other stimuli. An event with high predictive power will command access to the learning mechanism and thereby exclude the processing of other events. Although this basic idea has been developed by a number of theorists, it has reached its most sophisticated form in the attentional theory of Sutherland and Mackintosh (1971). The application of this form of limited-capacity theory assumes that associating the pretrained E1 with E2 during the first stage of a blocking experiment establishes this pretrained E1 as a predictor of an event of significance to the animal. This in turn ensures that pretrained E1 commands access to the limited-capacity learning mechanism. When the pretrained and added E1s are presented in compound during the second stage, processing of the pretrained element will have priority, and thus detract from learning about relationships involving the added E1.

Although this type of theory has had considerable success, especially when explaining phenomena concerned with discrimination learning (see Sutherland & Mackintosh, 1971), it is not clear how adequately it handles the simple forms of associative learning which we have been studying in this volume. To support this contention, let us look at yet another variant of the basic blocking experiment. Suppose that in the first stage we expose hungry rats to pairings of a light and food followed in the second stage by pairings of a compound of the light and a novel tone stimulus with shock. Pairing the light with food in the first stage should establish the light as a predictor of a significant event, the delivery of food, so that on compound trials the central learning mechanism should be primarily devoted to processing the light at the expense of the tone. Consequently, when we test for learning about the tone → shock relationship by presenting the tone alone, we should expect to observe less learning in this blocking group than in a control group for which the light was not established as a predictor of food in the first stage.

In fact this type of transreinforcer blocking does not appear to occur. Recently I performed just such an experiment (Dickinson, 1977) in which the amount learned about the tone → shock association was measured by observing the extent to which presentations of the tone suppressed lever-pressing for food by rats after compound training. The paired group received pairings of the light and food in Stage 1, whereas for the random control group these two events were uncorrelated. Rather than blocking learning about the tone → shock association, exposure to the light → food association in Stage 1 enhanced such learning. This is not the place to go into the reasons for such enhancement (see Dickinson & Pearce, 1977; Dickinson & Dearing, 1979); however, it is clear that the failure to find transreinforcer blocking provides difficulties for a simple limited-capacity model of the processing of E1.

Predictive power–The Mackintosh theory

The basic notion we are trying to capture is the idea that the processing of some target E1 depends on the degree to which it is a valid predictor of E2 relative to other E1s. Recently Mackintosh (1975a) has attempted to formalize this notion. According to Mackintosh, following every learning trial each of the E1s present is evaluated to see whether it is a good predictor of the E2 that occurred on that trial. This information is then used to control the processing of each E1 during subsequent learning experiences. If an E1 is a better predictor of E2 than all the other events present on that trial, the degree of processing which this E1 receives on a later trial will be increased, whereas if it is no better than the other events subsequent processing will be reduced. As well as learning about the relationships between events, the animal also learns whether or not particular events are good predictors and this learning then controls the degree to which these events are processed subsequently by the mechanism responsible for setting up the associative representations.

Throughout this book I have emphasized that successful associative learning depends upon the presentation of a surprising or unexpected E2. It is important to realize that Mackintosh's theory attributes a very different role to a surprising E2 to that espoused by Rescorla and Wagner. This difference can be illustrated by considering the first compound trial of a blocking experiment. On this first compound trial the presentation of E2 should not be surprising as it is predicted by the pretrained E1. According to Rescorla and Wagner blocking should occur on this first trial because E2, being expected, will not receive the necessary processing for the animal to learn about the association between the added E1 and E2. By contrast, Mackintosh argues that the level of learning is controlled by variations in the processing of E1, not E2. Furthermore, as the animal has not had the opportunity to evaluate the relative predictive power of the pretrained and added E1s prior to the first compound trial, learning should proceed normally on this first trial. Following this first compound trial, however, the animal will have had an opportunity to find out that the added E1 is in fact a worse predictor than the pretrained E1 of the E2 which actually occurred on that trial, and consequently it will be able to decrease the amount of processing received by the added E1 on subsequent trials. Blocking occurs, according to this account, because the animal learns that the added E1 is a relatively poor predictor of E2 during the initial compound trials and therefore fails to learn about the association on later trials.

One obvious prediction follows from Mackintosh's theory; blocking should not be observed with only a single compound trial, for prior to this trial the animal would not yet have found out that the added E1 is a redundant predictor. The evidence on this point is somewhat equivocal. Mackintosh (1975b) himself has consistently failed to find blocking following a single compound trial using a conditioned suppression procedure, whereas both Revusky (1971) and Gillan and Domjan (1977) have reported such blocking with a taste-aversion conditioning paradigm. However, more compelling

Table 12. *Design of the Mackintosh, Bygrave and Picton (1977) experiment*

	Stage 1	Stage 2		
Groups	(trial 4)	Trial 1	Trial 2	Test
0	L → sh	TL → sh		T
sh	L → sh	TL → sh-10-sh		T
sh-0	L → sh	TL → sh-10-sh	TL → sh	T
0-0	L → sh	TL → sh	TL → sh	T
0-sh	L → sh	TL → sh	TL → sh-10-sh	T

T: tone stimulus; L: light stimulus; sh: shock; sh-10-sh: shock followed by a second post-trial shock 10 seconds later.

evidence for the type of mechanism envisaged by Mackintosh comes from studies of the action of post-trial surprise. Recall that Dickinson *et al*. (1976) found that the unexpected presentation of a post-trial shock shortly after each compound trial enhanced learning to the added E1 (see p. 140). I pointed out that this result was compatible with the Rescorla–Wagner model; the post-trial shock, being unpredicted by the pretrained E1, should receive adequate processing for the animal to learn about the relationship between it and the added E1. Mackintosh's theory, however, provides an alternative explanation. Presentation of the post-trial shock on the initial compound trials ensured that the added E1, although a relatively poor predictor of the first shock, was as good a predictor as the pretrained E1 of another event of importance, namely the post-trial shock. As a result, the added E1 should have received sufficient processing on subsequent compound trials for learning to occur.

Clearly the two theories provide very different descriptions of how post-trial surprise works; Rescorla and Wagner assume that it serves to enhance learning on the immediately preceding trial by affecting E2 processing, whereas for Mackintosh it enhances learning on subsequent trials by maintaining the processing of the added E1. Given these different predictions, Mackintosh, Bygrave and Picton (1977) attempted to find out whether post-trial surprise augmented learning on the immediately preceding trial or on subsequent trials. Their experiment is complex and the rationale is best discussed by reference to table 12 which illustrates the basic design. In Stage 1 all five groups received four trials on each of which the presentation of a light was terminated with a shock. Then in Stage 2 a tone was compounded with the light and the compound was paired with a shock. Either one or two trials were presented in this second stage depending upon the group. One trial was given per day. Finally, the amount that the rats had learned about the tone → shock association was measured by presenting the tone by itself to see how much it suppressed drinking.

One pair of groups just received a single compound trial in Stage 2. For one group, Group 0, the tone–light compound was followed only by a single shock which was already predicted by the light on the basis of Stage 1 training. Group sh received exactly the same schedule except that a second post-trial shock was delivered 10 seconds after the first. If we focus solely on the role of E1 processing, Mackintosh would anticipate that these two groups would show the same amount of learning about the tone → first shock relationship. As the tone was a novel event to both groups on the compound trial, it should have been processed to the same extent by all animals with the level of processing being determined entirely by factors such as its salience. As a single compound trial was given, there was no opportunity for the processing of the tone to decrease prior to the pairing of the tone and first shock. Of course, Group sh may have also learned about the relationship between the tone and the second, post-trial shock. Figure 35 shows that in fact there was no major difference between the amount the animals in Groups 0 and sh learned about the tone → shock relationship. The level of suppression maintained by the tone was the same in both groups showing that surprising post-trial events do not significantly enhance learning on the immediately preceding trial as Rescorla and Wagner might anticipate.

The fact that the post-trial shock did not affect learning about the tone → shock association on the preceding trial does not mean that it was without any effect. In Group 0 the tone was presented in the absence of any surprising event, and Mackintosh's theory argues that subsequent processing of the tone should have declined as a result of this exposure. This means that if we gave a further tone–shock pairing, learning should be retarded on this second trial; Group 0-0 received just such a second trial. On the other hand, in Group sh the presentation of the tone was followed by a surprising post-trial event, and if such events tend to prevent the loss of processing, we should expect animals in this condition to show significantly more learning on a subsequent trial. Group sh-0 received a second compound trial following a post-trial shock on the first, and a comparison of the suppression to the tone in Groups sh-0 and 0-0 should tell us whether

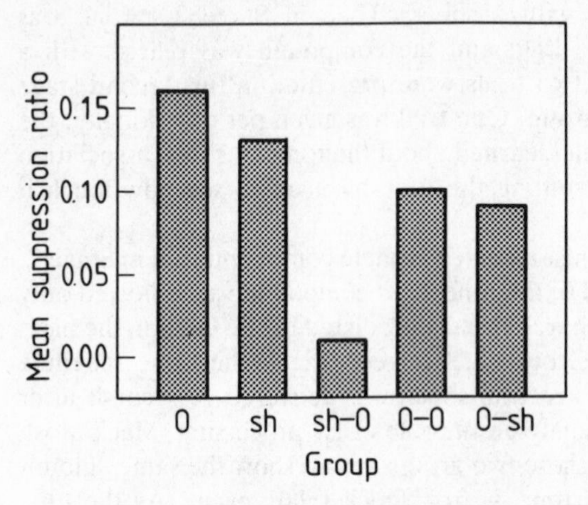

Fig. 35. The degree to which a tone suppressed licking for water after rats had been exposed to the blocking procedure outlined in table 12. Groups 0 and sh received only a single tone–light compound trial in Stage 2 which was followed by a post-trial shock in Group sh. The remaining groups received two compound trials in Stage 2. A post-trial shock was delivered following the first compound trial in Group sh-0, following the second in Group 0-sh, and following neither in Group 0-0. (After Mackintosh *et al.*, 1977.)

presence of the post-trial shock on the first compound trial enhanced learning on the second. Clearly it did, for as figure 35 shows, the tone produced more suppression on test in Group sh-0 than Group 0-0.

This pattern of results is entirely compatible with the idea that presenting the tone unaccompanied by a surprising event leads to a decrease in the ability of the animal to learn about relationships involving the tone on subsequent trials. Following the first compound trial by a surprising post-trial shock maintains the processing of the tone, so that the animal can learn about the relationship between it and the shock on the second compound trial. If the surprising post-trial shock acts in this pro-active manner, scheduling such an event on the second of two compound trials should be ineffective in enhancing learning about the tone. In a final group, Group 0-sh, Mackintosh *et al*. gave two compound trials, but in this case delivered the post-trial shock only after the second one (see table 12). As figure 35 illustrates, this group showed no more suppression to the tone than Group 0-0 which did not experience a surprising post-trial event on either trial.

These interesting results provide compelling support for

Mackintosh's account of why learning depends upon the surprising nature of E2. The presentation of an unexpected E2 may not only affect processing of this event itself, but also serve to maintain the effectiveness of any E1 associated with it. Essentially the same explanation can be given of why the unexpected omission, as well as addition, of a post-trial shock attenuates blocking, although we shall not develop the argument here (see Dickinson & Mackintosh, 1979).

I am now in a position to present a formal statement of Mackintosh's ideas. Such a formalization can be expressed within the framework of the general model of learning in which the change in the associative strength of an E1 depends upon the conjoint processing of this E1 and the E2. We have already met the idea that the processing of an E1 reflects its salience, which was designated by the parameter α. The core of Mackintosh's model is the idea that the processing of E1 can change with experience, and he expresses this idea by formulating rules for changing the effective salience of E1. As the changes in the processing of this event depend upon its relative predictive power, we need some way of stating the predictive power of an event. I have already pointed out that an E1 can be regarded as a perfect predictor of an E2 if the associative strength of this E1 is equal to the asymptote of the associative strength for E2 (λ), or, in terms of our processing model, the amount of processing received by E2 when it is totally unexpected. To the extent that there is a discrepancy between this asymptote and the associative strength of E1, E1 is not a perfect predictor of E2. Thus the degree to which some target E1, event A say, predicts E2 can be expressed by the discrepancy $|\lambda - V_A|$ where V_A is the associative strength of event A. The larger the value of this discrepancy the lower is the predictive power of A. It is important to realize, however, that it is the absolute value of this discrepancy, or in other words its modulus, which is important. If the associative strength of a particular E1 is either larger or smaller than the asymptote then this E1 is not a perfect predictor of E2. Furthermore it is not simply the absolute predictive power of A which is important in determining its subsequent processing, but rather its predictive power relative to all the other potential E1s present. The predictive power of these other E1s can be similarly expressed by a discrepancy of the form $|\lambda - V_X|$ where V_X is the combined associative strengths of all the other E1s present except for event A.

By comparing the size of the discrepancy for event A with that for all the other events present, we can express the relative predictive power of event A. Remember that the central tenet of Mackintosh's theory is that the subsequent processing of A, or in other words its effective salience, α_A, increases following a trial when A is a better predictor of E2 than all the other stimuli and decreases when A is not a better predictor. This idea can be expressed by the following rule for the change in α_A on a trial:

$$d\alpha_A > 0 \text{ if } |\lambda - V_A| < |\lambda - V_X| \qquad 10$$
$$d\alpha_A < 0 \text{ if } |\lambda - V_A| \geqslant |\lambda - V_X|$$

The size of the change in α_A on each trial is proportional to the magnitude of the inequalities in equation 10. Thus equation 10 specifies the way in which the processing of an E1 varies.

So far Mackintosh's model is only partially complete, and we have yet to address the way in which the processing of E2 also affects changes of associative strength. Rescorla and Wagner argued that the processing of E2 varies during the course of learning depending upon the extent to which E2 is predicted. Such variation accounts for the type of selective learning seen in blocking and related phenomena. By contrast, Mackintosh does not need to appeal to variations in E2 processing; the whole burden of selective learning can be attributed to changes in the processing of E1 brought about by the application of equation 10. Consequently one solution would be to assume that E2 processing is constant and simply determined by the salience and magnitude of this event, so that an increment in associative strength is simply a product of the current level of E1 processing and the fixed value of E2 processing. The problem with this idea is that learning should never stop if the processing of E1 is maintained, and yet we have seen that most behavioural indices of learning reach a maximum value or asymptote which cannot be increased by further pairings. In fact Mackintosh suggests that changes in the associative strength of a target E1, event A, are governed by the following equation:

$$dV_A = \alpha_A \beta(\lambda - V_A) \qquad 11$$

where α_A reflects the processing of the target E1, event A, and $\beta(\lambda - V_A)$ the processing of E2. According to equation 11, the animal will learn no more about the relationship between A and E2 from pairings of the two events once the associative strength of A equals the asymptote for E2. Unlike the Rescorla–Wagner model (see equation 8), equation 11 assumes that the presence of other E1s has no effect on E2 processing and the increment in the associative strength of the target E1 just depends upon its own current associative strength. From our point of view, the problem with equation 11 is that the discrepancy ($\lambda - V_A$) does not have any obvious psychological interpretation within the terms of our general processing model. For the Rescorla–Wagner theory the idea that an increment in learning is a function of the discrepancy between the asymptote and current associative strengths of all the E1s present captures the central notion of this theory about the role of surprise in learning. In Mackintosh's model, however, surprise does not act via a comparable discrepancy and its role within equation 11 is not readily interpretable in terms of processing mechanisms. I should point out, however, that Mackintosh (1975a) did not present his theory in the context of the type of processing model we have developed, and equation 11 can probably receive a psychological interpretation within a different framework.

Controlled and automatic processing–The Pearce–Hall theory

It could be argued that there is something intuitively implausible about the central idea of Mackintosh's theory, the idea that animals learn about an event to the extent that it has been a reliable and good predictor in the past. Certainly an animal should control its behaviour on the basis of the information provided by such reliable predictors. It is far less clear, however, that the learning capacity of an animal should be largely devoted to processing events which in the animal's recent history have been constituents of stable relationships. Rather one might expect the animal to devote most of its processing capacity to analysing events whose predictive significance is uncertain in an attempt to discover relationships involving these events.

This latter idea can be stated in terms of a distinction between automatic and controlled processing which has emerged in recent discussions of human information processing. For instance, Shiffrin and Schneider (1977) have suggested that the presentation of a stimulus can be processed in one of these two modes. When an event has had a consistent predictive significance in the past, it is processed in an automatic mode which allows the subject to take the appropriate action on the basis of what he has already learned about relationships involving this event. In terms of our general perspective, this means that an E1 with high predictive significance can readily engage the type of knowledge–action translation processes we discussed in the last chapter. However, this type of automatic processing does not allow the subject to learn about new relationships involving this event, and so cannot bring about changes in the associative representations encoding these relationships. For this to happen, the event must be processed in a controlled mode, and such processing occurs when the event is not a consistent predictor.

This view of learning has one clear implication that is at variance with Mackintosh's theory. If we establish an E1 as a good predictor by consistently pairing it with some E2, then the rate at which an animal will learn about a new relationship involving this E1 and a second, novel E2 should be retarded. Once the E1 is established as a good predictor of the first E2, it will be no longer processed in the controlled mode and so the animal will have initial difficulty in learning about an association between it and a new E2. We have, in fact, already discussed such an experiment. Recall that I (Dickinson, 1976) gave two groups of rats a number of pre-exposures to a tone followed by experience of a tone → shock association (see p. 144). For the paired group, presentations of the tone during pre-exposure were paired with food, whereas the random group received uncorrelated presentation of the tone and food. When we discussed this experiment previously, we emphasized the fact that the paired group showed more rapid learning about the tone → shock association than the random group, indicating that establishing the tone as a predictor during pre-exposure appeared to maintain the extent to which it was subsequently processed. However, it was also notable that the paired group learned about the tone → shock association more slowly than a group never pre-exposed to the tone, the food-alone group (see figure 34). Thus it appears that establishing the tone as a good predictor, in this case for food, although attenuating the decline in processing produced by simple pre-exposure, does not abolish the complete decrement. In fact, the rats actually learn slower when the tone is a good predictor than when it is completely novel.

This effect is demonstrated even more dramatically by a recent experiment reported by Hall and Pearce (1979). In the first stage one group of rats, the tone-alone group, simply received a number of exposures to a tone, whereas two other groups, the light–shock and tone–shock groups, received a number of pairings of a light and mild shock and tone and mild shock respectively. These stimulus–shock pairings produced some suppression of lever-pressing for food. In the second stage, all groups experienced pairings of a tone and stronger shock, and the rate at which the tone came to suppress lever - pressing for food was measured. Figure 36 shows that a basic latent inhibition effect emerged in that the light–shock group, which was never pre-exposed to the tone, learned faster than the tone-alone group. The really interesting result, however, is the rate at which the tone–shock group learned in the second stage. If, as Mackintosh argues, the processing of an E1 is increased by making it a predictive

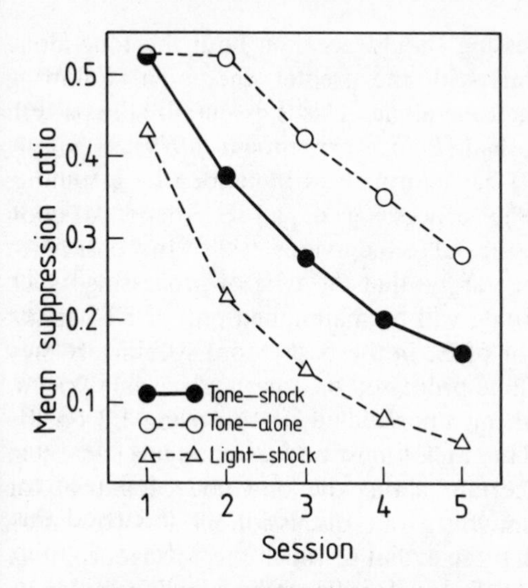

Fig. 36. The suppression of lever-pressing for food by rats produced by a tone during a series of sessions in which the tone was paired with shock. Previously the tone–shock and light–shock groups had received pairings of the tone and light respectively with a weaker shock, whereas the tone-alone group had been simply pre-exposed to the tone. (After Hall & Pearce, 1979.)

stimulus, the tone–shock group should learn at least as rapidly as the light–shock group. The experience of the relationship between the tone and weak shock in the first stage should have established the tone as a good predictor of an event of significance, which at the very least should have prevented any decline in the processing of the tone and possibly even increased its effectiveness. In fact, as figure 36 shows, the rats in the tone–shock group learned more slowly in the second stage than those in the light–shock group.

This pattern of results is compatible with the idea that E1 loses its capacity to be processed to the extent to which it is a good predictor. In the tone-alone condition the tone initially predicts nothing and nothing occurs so that the tone is a valid predictor and its effectiveness should rapidly decline. In the tone–shock group, on the other hand, the tone initially predicts nothing, but a weak shock occurs. Consequently, during the initial trials of the first stage the tone is a poor predictor, and its effectiveness will either be maintained or actually increase. However, once the animal has learned about the tone → weak shock association, the tone will have been established as a good predictor and its capacity for processing will be attenuated.

Thus a decrease in processing should occur in both the tone-alone and tone–shock conditions with the greater decrement following simple pre-exposure to the tone alone. This is essentially the pattern of results revealed by the Hall–Pearce experiment (1979).

Pearce and Hall (1980) have formalized their idea by assuming, like Mackintosh, that the processing of an E1 depends upon the predictive power of the E1 on previous trials. In contrast to Mackintosh, however, they argue that the type of processing which results in associative learning will be maintained only if E1 has not been a successful predictor of E2 in the past. More specifically they argue that a target E1 will be processed on a particular trial, Trial n, only if it was paired with an unpredicted or surprising E2 on the previous trial, Trial n-1. Thus an E1 must be presented in a context in which the animal is uncertain about the outcome of a trial for processing to be maintained. In our discussion of the Rescorla–Wagner model, we saw that the extent to which the occurrence of an E2 on a particular trial, say Trial n-1, is unpredicted will be given by the discrepancy $|\lambda^{n-1} - \Sigma V^{n-1}|$. The term λ^{n-1} is the asymptote of the associative strength for the E2 presented on Trial n-1 and ΣV^{n-1} is the sum of the associative strengths of all the E1s present on Trial n-1 for this E2. If the discrepancy $|\lambda^{n-1} - \Sigma V^{n-1}|$ is large, the occurrence of E2 is surprising and so the processing of the target E1 should be high on the next trial, Trial n. If we take α^n to be the processing of the target E1 on Trial n, we can state the Pearce–Hall rule for determining the amount of processing of E1 as follows:

$$\alpha^n = |\lambda^{n-1} - \Sigma V^{n-1}| \qquad 12$$

Equation 12 simply expresses the idea that if E1 is paired with a surprising E2 on Trial n-1, it will be effectively processed on Trial n, whereas if it is paired with a predicted E2, there will be little or no processing on the subsequent trial.

As in Mackintosh's theory, the whole burden of selective learning is placed on variations in E1 processing; for instance, blocking occurs because presentation of the pretrained E1 ensures that E2 is fully predicted on the initial compound trials so that the processing of the added E1 rapidly declines during compound training. In addition, an elegant feature of the Pearce–Hall theory lies in the fact that it does not need to assume that E2 processing changes at all during the course of learning. In the case where the animal is exposed to simple E1–E2 pairings, no further increments in learning will occur as soon as E1 perfectly predicts E2 because, when this point has been reached, E1 will no longer receive the processing necessary for associative learning. Of course, E1 will still be processed in a mode that allows its presentation to activate the necessary knowledge–action translation systems for producing the appropriate behaviour. Pearce and Hall argue that a particular E2 is always processed to the same degree which is simply determined by its magnitude (λ). As a result, we can use our basic model of learning to state a function for changing the associative strength of a particular E1 in the following form:

$$dV = \alpha S \lambda \qquad 13$$

where αS reflects the processing of E1 and λ that of E2. Pearce and Hall include the parameter S to specify the way in which the salience of E1 affects learning. They choose to separate this factor explicitly from the parameter α which is taken as a measure of what they call the associability of E1. The salience is determined by the physical properties of E1 in relation to the animal's sensory systems, whereas the associability is determined by the past predictive power of E1.

This concludes our discussion of the various accounts of why animals only show sustained learning when E2 is surprising. As we have seen, at least three distinct types of mechanism have been suggested over the last decade, and at present none of them have received universal acceptance. In fact the empirical evaluation of these various theories represents one of the main strands of current research in the area. Our discussion of this research has been restricted so far to cases of E1 → E2 learning, and now we must extend our treatment to the other half of associative learning, namely that of E1 → no E2 learning.

E1 → no E2 learning

One of the attractive features of the original Rescorla–Wagner theory is that it provides a simple and straightforward account of the mechanism controlling E1 → no E2 learning. To illustrate the application of the model, let us consider the experiment by Rescorla (1973) discussed in chapter 2. Using a conditioned suppression procedure with rats, Rescorla presented the animals with a series of trials in which a light was paired with a shock intermixed with trials in which a compound of the light and tone was presented in the absence of any shock. Thus the tone was paired with the non-occurrence of shock at a time when the animal would expect the shock due to the presence of the light. Following this training, the light suppressed lever-pressing for food, indicating that the animals had learned about the light → shock association. In addition, the tone acquired the ability to inhibit the suppression produced by the light, thus providing a behavioural index that the rats had also learned about the tone → no shock relationship.

The question is whether we require any further principles, beyond those we have discussed for E1 → E2 learning, in order to be able to understand how the animal came to learn about this tone → no shock association. Let us approach this problem by simply trying to apply the Rescorla–Wagner model, as specified by equation 8, on the tone–light compound trials to determine the change in the associative strength of the tone. The change in the associative strength of the tone on a tone–light compound is given by:

$$dV_T = \alpha_T\beta(\lambda - \Sigma V)$$

The term ΣV refers to the sum of the associative strengths of all the E1s present on this trial which in this case would have been the sum of the associative strengths of the light and the tone. On the initial compound trial the associative strength of the tone would have been zero, whereas that for the light should have had a positive value as the light was independently paired with the shock. The asymptote of the associative strength for the shock, λ, being governed by its intensity, would have been zero as no shock was presented on compound trials. Thus the change in the associative strength of tone should be given by:

$$\begin{aligned} dV_T &= \alpha_T\beta(\lambda - (V_T + V_L)) \\ &= \alpha_T\beta(0 - (0 + V_L)) \\ &= -\alpha_T\beta V_L \end{aligned}$$

Thus we see that the change in the associative strength brought about by a tone–light compound trial would have been negative, and in fact the associative strength of the tone should have become progressively more negative across successive compound trials until it equalled the opposite value to that of the light. This means, of course, that when the tone was combined with the light, the total associative strength of the compound should have been less than that of the light alone. As a result the tone had the capacity to inhibit the level of suppression elicited by the light.

According to the Rescorla–Wagner theory learning about an E1 → no E2 association consists of the acquisition of a negative associative strength. The great virtue of the theory is the simple and unified account it provides of both E1 → E2 and E1 → no E2 learning which can be encompassed by a single equation. There are two major problems, however, one empirical and the other theoretical. As it stands, the theory predicts changes in the inhibitory properties of an event in cases where they do not seem to occur. This problem can be illustrated by an experiment of Zimmer-Hart and Rescorla (1974) on the extinction of inhibition. The question addressed by this experiment is whether presenting E1 alone after exposure to an E1 → no E2 relationship would abolish the inhibitory properties of E1. As in the previous conditioned suppression experiment, the rats received a series of trials in which a light was paired with shock intermixed with trials on which a tone–light compound was presented alone. On half of these compound trials, the tone was of a high frequency and on the remaining trials of a low frequency. This training schedule led to rats suppressing responding in the presence of the light alone, but not in the presence of the tone–light compounds. This indicated that the tones had acquired the capacity to inhibit the suppresion elicited by the light, or in terms of the Rescorla–Wagner theory negative associative strength.

All the animals then received 72 presentations of one of the tones by itself. We can find out what effect each of these presentations should have had on the associative strength of this tone by applying the Rescorla–Wagner equation on one of these trials. The asymptote should have been zero because no shock was presented, while the associative strength of the tone should have had a negative value as a result of the initial training. Thus the change in the associative strength of the tone on a trial on which it is presented alone is given by:

$$\begin{aligned} dV_T &= \alpha_T\beta(0 - (-V_T)) \\ &= \alpha_T\beta V_T \end{aligned}$$

Each simple exposure of the tone should have resulted in a positive increment in its associative strength which should have gradually erased its initial negative associative strength and thus its capacity to act as an inhibitor. Of course, the other non-exposed tone should have retained its full negative associative strength and its inhibitory capacity. To see whether these changes had occurred, Zimmer-Hart and Rescorla gave test presentations of the light alone and compounds of the light plus each tone in turn. The light suppressed responding and both exposed and non-exposed tones inhibited suppression to the same extent. The inhibitory capacity of the exposed tone was completely unaffected by the 72 presentations. This experiment was but one of a number of attempts by Zimmer-Hart and

Rescorla to change the inhibitory properties of a stimulus by simple exposure, all of which were without effect. As a result, these experiments expose a clear limitation on the ability of the Rescorla–Wagner theory to handle E1 → no E2 learning.

The theoretical problem with the model comes when we try to construct an internal representation of an E1 → no E2 association which is congruent with the concept of a negative associative strength. In the last chapter we argued that learning about an E1–E2 association consists of setting up some form of internal representation of this relationship. This perspective enables us to identify an associative strength of E1 as a measure of the strength of the underlying representation of an E1–E2 relationship. Given this perspective, however, it is not obvious how we can give a meaningful interpretation to the concept of a negative associative strength. The internal representation of an E1–E2 relationship can be either non-existent, in which case the relevant associative strength will be zero, or present within the animal's mental apparatus, in which case the associative strength will be positive. The psychological ambiguity inherent in the concept of a negative associative strength suggests that we should approach the problem of E → no E2 learning from a new perspective.

In the last chapter I argued that exposure to an E1 → no E2 relationship sets up a representation which is essentially independent of that established by an E1 → E2 association. If this is true, clearly we should identify two separate types of associative strength, one relating to the internal representation of an E1 → E2 association and the other to the representation of an E1 → no E2 relationship. We have spent most of this chapter discussing the processes governing changes in the associative strength of an E1 → E2 relationship, so let us now try to develop an analogous set of rules for changing that of an E1 → no E2 association. Of course, we could go through the laborious task of working out the rules for each type of learning model, but such an exercise would be redundant. Rather I shall illustrate the processes governing E1 → no E2 learning in terms of the Pearce–Hall theory; the basic rationale, however, could be equally well applied to Mackintosh's model or even a modified form of the Rescorla–Wagner theory.

The first problem we have to consider is the nature of the rules governing the processing of E1. Remember that the central idea of the Hall–Pearce theory is that an E1 will be processed to the extent that it has been recently paired with a surprising or unexpected event. This idea is embodied within equation 12. In order to see whether this equation has to be modified to take account of E1 → no E2 learning, let us consider an imaginary experiment, like that of Rescorla (1973), in which light–shock pairings are followed by presentations of a light–tone compound in the absence of shock. The processing of the tone after the first tone–light compound trial, Trial 1, can be derived by making the appropriate substitutions in equation 12. As no shock occurs, the asymptote of associative strength (λ) will be zero, as will the associative strength of the tone for the shock. If we assume that the animal has previously experienced pairings of the light and shock, the associative strength of the light for the shock will be positive. As a result, the processing of the tone on the next compound trial, Trial 2, will be given by:

$$\begin{aligned} \alpha_T^2 &= |\lambda^1 - \Sigma V^1| \\ &= |\lambda^1 - (V_T^1 + V_L^1)| \\ &= |0 - (0 + V_L^1)| \\ &= V_L^1 \end{aligned}$$

Thus on the second compound trial the extent to which the tone will be processed will depend upon the associative strength of the light for the shock on the first compound trial, or in other words the extent to which the animal expected the shock on this previous trial. This result makes psychological sense, for if the animal expected the shock, the tone would have been paired with a surprising event, the omission of the shock, and therefore should be processed on its next presentation. But what would have happened if the first compound trial also contained a third stimulus, a clicker for example, when the animal had previously experienced a clicker → no shock association? Now the omission of the shock on the first compound trial would have been fully predicted, so that the tone would have been paired with an entirely expected state of affairs. Under these circumstances, the subsequent processing of the tone should be low. Clearly we must take account of the presence of a predictor of the absence of E2 in assessing changes in the processing of a target E1. This can be done by modifying equation 12 in the following way:

$$\alpha^n = |\lambda^{n-1} - \Sigma V^{n-1} + \Sigma \overline{V}^{n-1}| \qquad 14$$

where $\overline{V}$ is the sum of the associative strengths of all E1s present for *no* E2 which reflects the strength of all the relevant E1 → no E2 representations. Application of equation 14 to our example in which a compound of a tone, light, and clicker is presented gives:

$$\alpha_T^2 = |\lambda_1 - (V_T^1 + V_L^1) + \overline{V}_C^1|$$
$$= |\overline{V}_C^1 - V_L^1|$$

where $\overline{V}_C$ is the associative strength of the clicker for no shock. If the associative strength of the light for the shock and the clicker for no shock are approximately equal so that the non-occurrence of the shock is fully predicted, α_T^2 will be approximately zero and the tone will receive little or no processing on subsequent trials. This means that the presence of the clicker, if it has been previously established as a predictor of no shock, could lead to a reduction of processing of the tone. The animals will then learn less about the tone → no shock relationship on subsequent compound trials. A little reflection will show that this is an example of the blocking of E1 → no E2 learning, which has been empirically demonstrated (Suiter & LoLordo, 1971).

Having outlined the rules governing the processing of E1 during E1 → no E2 learning, we must now turn our attention to the mechanism controlling the processing of the event constituted by the omission of E2. Recall that Pearce and Hall argue that the processing of E2 during E1 → E2 learning does not vary and is simply a function of the intensity of E2. Correspondingly we might expect the processing of the omission of E2 is also determined by the intensity of that event. But what might we mean by talking about the intensity of the omission of E2? Remember that E1 → no E2 learning occurs when E2 is omitted in a context in which the animal, on the basis of the presence of an E1 previously paired with E2, expects E2. One view might be that the intensity of the omission of E2 is equivalent to the degree the animal expects E2 on that trial. The value of this expectation is equivalent to the sum of the associative strengths of all the E1s present for E2 minus the sum of the associative strengths of all the E1s present for no E2, or, in other words, the term $\Sigma V - \Sigma\overline{V}$. Having specified the intensity of the omission of E2, we are now in a position to determine how the processing of E1 and the omission of E2 interact to produce a change in the associative strength of E1 for the omission of E2. Remember that our basic model argues that the increment in the associative strength of E1 for, in this case, no E2 is just given by the product of the processing of these two events, so that:

$$d\overline{V} = \alpha S(\Sigma V - \Sigma\overline{V}) \qquad 15$$

where $d\overline{V}$ is the change in the associative strength of E1 for no E2. The parallel to equation 13, specifying E1 → E2 learning, is clear. In summary then, equation 13 would be applied when an E2 is presented whereas equation 15 is applied when no E2 is presented. The former case results in E1 → E2 learning and the latter E1 → no E2 learning. In both cases the processing of E1, as specified by α, is determined by equation 14.

The psychological implications of this view of E1 → no E2 learning are radically different from those of the Rescorla–Wagner model. According to the Rescorla–Wagner model, the same associative strength variable underlies both E1 → E2 and E1 → no E2 learning. This variable can take both positive and negative values, and increases and decreases are controlled by the single Rescorla–Wagner function. Clearly this theory points to the idea that some unitary type of underlying associative representation is set up by both E1 → E2 and E1 → no E2 associations. All that differs is the strength of the representation in the two cases. Built into this model is the idea that unlearning can occur. If an animal is initially exposed to an E1 → E2 relationship a positive strength will develop. Subsequent presentations of an E1 → no E2 association will serve to decrease this associative strength, and thus wipe out the initial internal representation set up by exposure to the E1 → E2 relationship.

By contrast, the Pearce–Hall theory argues that exposure to E1 → E2 and E1 → no E2 associations set up independent representations, each with its own associative strength. These associative strengths can only take positive values, and only increase in magnitude. Exposure to an E1 → no E2 association does not produce unlearning in the sense of weakening the representation of an E1 → E2 association, but rather results in the animal learning something new. Of course, the knowledge–action translation systems engaged by a particular E1 will have to take into account both E1 → E2 and E1 → no E2 representations of which the event is a constituent in determining the animal's behaviour. This view of the mechanisms governing E1 → no E2 learning is clearly more in line with the theories of associative representations we discussed in chapter 3.

Process learning

Two features of associative learning lie outside my analysis of the learning mechanism so far: learned irrelevance and the role of the biological or causal relevance of E1 to E2. In chapter 2 I described experiments indicating that when an animal is exposed to an uncorrelated schedule of E1 and E2 presentations, subsequent learning about both E1 → E2 and E1 → no E2 associations is retarded. This is true whether E1 is either a stimulus or an action. It appears that animals can learn that two events are causally unrelated or irrelevant to each other. One way of accounting for this type of learning is to assume that an associative representation of the random relationship is formed, an idea touched on at the end of chapter 3. At the time, however, I pointed out the difficulty of characterizing such a representation, at least in terms of an excitatory-link model. In fact certain authors (e.g. Maier & Seligman, 1976) have taken this type of learning as evidence for a propositional form of representation.

An alternative account argues that learned irrelevance is not mediated by a change in the representational structures of the animal's memory, but reflects a change in the learning mechanism itself. In chapter 2 I pointed out that an uncorrelated E1–E2 schedule is effectively equivalent to the blocking procedure; in fact it is this equivalence which justifies the analytic priority we have accorded to the blocking effect throughout this book. This equivalence means that the changes in processing we observe during blocking should also occur during exposure to an uncorrelated schedule. According to the Rescorla–Wagner theory exposure to a blocking procedure should leave E1 essentially neutral or with a small positive associative strength. All the blocking procedure does is to ensure that changes in the associative strength of E1, which would otherwise have occurred, are minimized. There is no reason to expect retarded learning following exposure to either a blocking procedure or an uncorrelated schedule.

The Mackintosh and Pearce–Hall theories, on the other hand, argue that exposure to a blocking procedure does produce sustained changes in the processing of E1. The capacity of E1 to be effectively processed progressively decreases across the series of compound trials of the blocking procedure. The same decline, of course, should be observed during exposure to an uncorrelated schedule, which in turn will result in the retardation of any further learning.

This brief discussion of learned irrelevance highlights an important implication of these theories. There appear to be two different types of learning. Initially we defined learning as the formation of some form of internal representation of a relationship or association between events in the animal's environment. This kind of learning, associative learning, is reflected by a change in the associative strength of E1. However, it is clear that the Mackintosh and Pearce–Hall models assume that another form of enduring change results from the exposure to event associations, namely variations in the processing of E1. For want of a better term, we can refer to this type of learning as process learning for it reflects a change in the way the learning mechanism handles information about the occurrence of events. Failures of associative learning seen in the blocking, overshadowing, and learned irrelevance experiments are really examples of such process learning.

The phenomenon of causal relevance, however, presents a more radical challenge to the type of theories we have been considering, and has important implications for our general model of the learning mechanism. Recall from chapter 2 that the basic finding is that the amount animals learn about an E1 → E2 association from a pairing of the two events depends upon the qualitative nature of these events; a rat more readily learns about a taste → illness association than a buzzer → illness relationship, whereas buzzer → shock learning occurs more rapidly than taste → shock learning. In terms of our processing model, the obvious implication is that the processing received by an event depends not only upon the properties of that event, reflecting its salience and past predictive history, but also upon the properties of the other event with which it is associated. In other words there appears to be an interaction between the processing of E1 and E2.

At present we know very little about the determinants of this interaction and none of the theories we have considered provide a ready account. On the one hand, the interaction of the processing of E1 and E2 may simply reflect an innate property of the learning mechanism of the particular species we are considering. Alternatively the interaction could be the outcome of the type of process learning underlying learned irrelevance (see Mackintosh, 1973; Dickinson & Mackintosh, 1979). During its development an animal will observe consistent correlations between certain types of events, say tastes and gastric consequences, and a consistent lack of correlation between other events, for example auditory and visual stimuli and gastric effects. This general and pervasive developmental experience might result in a form of process learning, similar to that we have already considered, which in turn will affect associative learning later in the animal's life. Thus, for instance, the way in which a taste will be processed will depend upon the predictive history of this particular taste and possibly other stimuli from the same modality with respect to the particular type of E2 with which the taste is associated. Unfortunately we have no direct evidence that such process learning underlies the selectivity seen in causal relevance experiments. It is an attractive idea, however, for then we can view the change in the associative representations as providing the animal with specific and detailed information about particular current relationships in its environment, while process learning will yield information about the general causal structure of its world.

Summary

In this chapter we have discussed the nature of the learning mechanism responsible for setting up and changing associative representations when an animal is exposed to appropriate E1–E2 relationships. The basic model assumed that changes in a representation following a learning experience depend upon the extent to which E1 and E2 are conjointly processed by the learning mechanism. In chapter 2 we saw that a critical condition for successful associative learning was that E2 should be surprising or unpredicted. Thus the task faced in the present chapter was that of explaining why E1 and E2 receive the appropriate processing for learning only when E2 is unpredicted.

There are two general answers to this question. The first argues that the amount of processing received by an event depends upon the extent to which the event itself is predicted. Using this basic assumption, the Rescorla–Wagner model focuses on variations in the processing of E2; to the extent that E2 is predicted the animal fails to process it in a way that leads to learning. This theory provides a ready explanation of overshadowing and blocking, but fails to account for the variations in the processing of E1 which appear to occur when the animal is simply exposed to this event prior to learning. Wagner accounted for this phenomenon by extending the basic principle of the Rescorla–Wagner model to variations in the processing of E1 as well as E2. Even so, certain features of the attenuation of blocking by post-trial surprise could not be readily incorporated within this framework and encourage us to look at alternative views of the learning mechanism.

The second class of theories argue that the processing of E1 depends upon whether or not it has been an effective predictor of events of importance in the animal's past experience. Two radically different models were considered. Mackintosh suggested that an E1 will be effectively processed if this event has been a relatively good predictor of other events in the past, whereas according to Pearce and Hall such processing will only occur if E1 has been recently associated with a surprising or unpredicted E2. In the case where a single E1 is associated with E2, the Pearce–Hall theory maintains that E1 will receive adequate processing only if it is not already a good predictor of E2. An important implication of both these accounts, however, is that exposure to event relationships can lead to two different forms of learning. The first consists of setting up and modifying associative representations of event relationships, whereas the second can be seen as enduring changes in the way the learning mechanism processes information about events in light of their past predictive significance. The phenomena of learned irrelevance and possibly causal relevance, which lie outside the scope of the Rescorla–Wagner model, find a potential explanation in terms of this form of process learning.

No one theory has, as yet, received universal acceptance and all have their particular strengths and weakness within the various domains of associative learning. As we have seen, however, the empirical foundations of these theories have largely emerged during the last decade or so, and if comparable research impetus is maintained for the next few years we might well expect to see a resolution or integration of these accounts before too long.

References

Anderson, J. R. 1976. *Language, Memory and Thought.* Hillsdale, N. J.: Lawrence Erlbaum Associates.

Atkinson, R. C. & Shiffrin, R. M. 1971. The control of short-term memory. *Scientific American*, **225**, 82–90.

Bakal, C. W., Johnson, R. D. & Rescorla, R. A. 1974. The effect of change in US quality on the blocking effect. *Pavlovian Journal of Biological Sciences,* **9,** 97–103.

Baker, A. G. & Mackintosh, N. J. 1977. Excitatory and inhibitory conditioning following uncorrelated presentations of the CS and US. *Animal Learning and Behavior,* **5,** 315–19.

Blanchard, R. & Honig, W. K. 1976. Surprise value of food determines its effectiveness as a reinforcer. *Journal of Experimental Psychology: Animal Behavior Processes*, **2**, 67–74.

Craik, F. I. M. & Lockhart, R. S. 1972. Levels of processing: A framework for memory research. *Journal of Verbal Learning and Verbal Behavior*, **11**, 671–84.

Daly, H. B. 1974. Reinforcing properties of escape from frustration aroused in various learning situations. In *The Psychology of Learning and Motivation*, G. H. Bower (ed), vol. 8, pp. 187–228. New York: Academic Press.

Dickinson, A. 1976. Appetitive-aversion interactions: Facilitation of aversive conditioning by prior appetitive training in the rat. *Animal Learning and Behavior*, **4**, 416–20.

Dickinson, A. 1977. Appetitive-aversive interactions: Superconditioning of fear by an appetitive CS. *Quarterly Journal of Experimental Psychology*, **29**, 71–83.

Dickinson, A. & Boakes, R. A. (eds) 1979. *Mechanisms of Learning and Motivation: A Memorial Volume to Jerzy Konorski*. Hillsdale, N. J.: Lawrence Erlbaum Associates.

Dickinson, A. & Dearing, M. F. 1979. Appetite-aversive interactions and inhibitory processes. In *Mechanisms of Learning and Motivation*, A. Dickinson and R. A. Boakes (eds), pp. 203–32, Hillsdale, N. J.: Lawrence Erlbaum Associates.

Dickinson, A., Hall, G. & Mackintosh, N. J. 1976. Surprise and the attenuation of blocking. *Journal of Experimental Psychology: Animal Behavior Processes*, **2**, 313–22.

Dickinson, A. & Mackintosh, N. J. 1979. Reinforcer specificity in the enhancement of conditioning by posttrial surprise. *Journal of Experimental Psychology: Animal Behavior Processes*, **5**, 162–77.

Dickinson, A. & Pearce, J. M. 1977. Inhibitory interactions between appetitive and aversive stimuli. *Psychological Bulletin*, **84**, 690–711.

Dinsmoor, J. A. 1978. Escape, avoidance, punishment: Where do we stand? *Journal of the Experimental Analysis of Behavior*, **28,** 83–95.

Domjan, M. & Wilson, N. E. 1972. Specificity of cue to consequence in aversion learning in the rat. *Psychonomic Science*, **26**, 143–5.

Dunham, P. J. 1971. Punishment: Method and theory. *Psychological Review*, **78**, 58–70.

Dweck, C. S. & Wagner, A. R. 1970. Situational cues and correlation between CS and US as determinant of the conditioned emotional response. *Psychonomic Science*, **18**, 145–7.

Fodor, J. A. 1977. *The Language of Thought*. Hassocks, Sussex: Harvester Press.

Garcia, J., Kimmeldorf, D. J. & Koelling, R. A. 1955. Conditioned aversion to saccharin resulting from exposure to gamma radiation. *Science*, **122**, 157–8.

Gibbon, J., Berryman, R. & Thompson, R. L. 1974. Contingency spaces and measures in classical and instrumental conditioning. *Journal of the Experimental Analysis of Behavior*, **21**, 585–605.

Gillan, D. J. & Domjan, M. 1977. Taste-aversion conditioning with expected versus unexpected drug treatment. *Journal of Experimental Psychology: Animal Behavior Processes*, **3**, 297–309.

Glazer, H. I. & Weiss, J. M. 1976. Long-term interference effects: An alternative to 'learned helplessness'. *Journal of Experimental Psychology: Animal Behavior Processes*, **2**, 202–13.

Goldman, A. I. 1970. *A Theory of Human Action*. Princeton, N. J.: Princeton University.

Gormezano, I. & Tait, R. W. 1976. The Pavlovian analysis of instrumental conditioning. *Pavlovian Journal of Biological Sciences,* **11,** 37–55.

Gustavson, C. R., Kelly, D. J., Sweeney, M. J. & Garcia, J. 1976. Prey–lithium aversion I: Coyotes and wolves. *Behavioral Biology*, **17**, 61–72.

Hall, G. & Pearce, J. M. 1979. Latent inhibition of a CS during CS–US pairings. *Journal of Experimental Psychology: Animal Behavior Processes*, **5,** 31–42.

Holland, P. C. 1977. Conditioned stimulus as a determinant of the form of the Pavlovian conditioned response. *Journal of Experimental Psychology: Animal Behavior Processes,* **3,** 77–104.

Holland, P. C. 1979. Differential effects of omission contingencies on various components of Pavlovian appetitive responding in rats. *Journal of Experimental Psychology: Animal Behavior Processes*, **5**, 178–93.

Holland, P. C. & Rescorla, R. A. 1975a. The effect of two ways of devaluing the unconditioned stimulus after first-and second-order appetitive conditioning. *Journal of Experimental Psychology: Animal Behavior Processes*, **1**, 355–63.

Holland, P. C. & Rescorla, R. A. 1975b. Second-order conditioning with food unconditioned stimuli. *Journal of Comparative and Physiological Psychology*, **88**, 459–67.

Holland, P. C. & Straub, J. J. 1979. Differential effects of two ways of devaluing the unconditioned stimulus after Pavlovian appetitive conditioning. *Journal of Experimental Psychology: Animal Behavior Processes,* **5**, 65–78.

Hyde, T. S. 1976. The effect of Pavlovian stimuli on the acquisition of a new response. *Learning and Motivation,* **7,** 223–39.

Jackson, R. L., Maier, S. F. & Rapaport, P. M. 1978. Exposure to inescapable shock produces both activity and associative deficits. *Learning and Motivation*, **9**, 69–98.

Jenkins, H. M. & Moore, B. R. 1973. The form of the auto-shaped response with food or water reinforcers. *Journal of the Experimental Analysis of Behavior*, **20**, 163–81.

Kamin, L. J. 1969. Predictability, surprise, attention and conditioning. In *Punishment and Aversive Behavior*, B. A. Campbell and R. M. Church (eds), pp. 279–96, New York: Appleton–Century–Crofts.

Konorski, J. 1948. *Conditioned Reflex and Neuron Organisation.* Cambridge: Cambridge University Press.

Konorski, J. 1967. *Integrative Activity of the Brain: An Interdisciplinary Approach*. Chicago: University of Chicago Press.

Leyland, C. M. 1977. Higher-order autoshaping. *Quarterly Journal of Experimental Psychology*, **29**, 607–19.

LoLordo, V. M. 1979. Selective associations. In *Mechanisms of Learning and Motivation*, A. Dickinson and R. A. Boakes (eds), pp. 367–99, Hillsdale, N. J.: Lawrence Erlbaum Associates.

Mackintosh, N. J. 1973. Stimulus Selection: Learning to ignore stimuli that predict no change in reinforcement. In *Constraints on Learning*, R. A. Hinde and J. Stevenson-Hinde (eds), pp. 75–96, London: Academic Press.

Mackintosh, N. J. 1974. *The Psychology of Animal Learning*. London: Academic Press.

Mackintosh, N. J. 1975a. A theory of attention: Variations in the associability of stimulus with reinforcement. *Psychological Review*, **82**, 276–98.

Mackintosh, N. J. 1975b. Blocking of conditioned suppression: Role of the first compound trial. *Journal of Experimental Psychology: Animal Behavior Processes*, **1**, 335–45.

Mackintosh, N. J. 1976. Overshadowing and stimulus intensity. *Animal Learning and Behavior*, **4**, 186–92.

Mackintosh, N. J., Bygrave, D. J. & Picton, B. M. B. 1977. Locus of the effect of a surprising reinforcer in the attenuation of blocking. *Quarterly Journal of Experimental Psychology*, **29**, 327–36.

Mackintosh, N. J. & Dickinson, A. 1979. Instrumental (Type 11) conditioning. In *Mechanisms of Learning and Motivation*, A. Dickinson and R. A. Boakes (eds), pp. 143–70, Hillsdale, N. J.: Lawrence Erlbaum Associates.

Mahoney, W. J. & Ayres, J. J. B. 1976. One-trial simultaneous and backward fear conditioning as reflected in conditioned suppression of licking in rats. *Animal Learning and Behavior*, **4**, 357–62.

Maier, S. F. & Seligman, M. E. P. 1976. Learned helplessness: Theory and evidence. *Journal of Experimental Psychology: General*, **105**, 3–46.

Marler, P. A. 1970. A comparative approach to vocal learning: Song development in white-crowned sparrows. *Journal of Comparative and Physiological Psychology Monograph*, vol. 71, no. 2, pt 2, pp. 1–25.

Morris, R. G. M. 1975. Preconditioning of reinforcing properties to an exteroceptive feedback stimulus. *Learning and Motivation*, **6**, 289–98.

Odling-Smee, F. J. 1975. Background stimuli and the inter-stimulus interval during Pavlovian conditioning. *Quarterly Journal of Experimental Psychology*, **27**, 387–92.

Pavlov, I. P. 1927. *Conditioned Reflexes*. Oxford: Oxford University Press.

Pearce, J. M. & Hall, G. 1980. A model for Pavlovian learning: Variations in the effectiveness of conditioned but not of unconditioned stimuli. *Psychological Review* (in press).

Premack, D. 1976. *Intelligence in Ape and Man*. Hillsdale, N. J.: Lawrence Erlbaum Associates.

Rashotte, M. E., Griffin, R. W. & Sisk, C. L. 1977. Second-order conditioning of the pigeon's key peck. *Animal Learning and Behavior*, **5**, 25–38.

Rescorla, R. A. 1967. Pavlovian conditioning and its proper control procedures. *Psychological Review*, **74**, 71–80.

Rescorla, R. A. 1968. Probability of shock in the presence and absence of CS in fear conditioning. *Journal of Comparative and Physiological Psychology*, **66**, 1–5.

Rescorla, R. A. 1969a. Conditioned inhibition of fear resulting from negative CS–US contingencies. *Journal of Comparative and Physiological Psychology*, **67**, 504–9.

Rescorla, R. A. 1969b. Establishment of a positive reinforcer through contrast with shock. *Journal of Comparative and Physiological Psychology*, **67**, 260–3.

Rescorla, R. A. 1971. Variations in the effectiveness of reinforcement and nonreinforcement following prior inhibitory conditioning. *Learning and Motivation*, **2**, 113–23.

Rescorla, R. A. 1973. Second-order conditioning: Implications of theories of learning. In *Contemporary Approaches to Conditioning and Learning*, F. J. MacGuigan and D.B. Lumsden (eds), pp. 127–50, Washington, D. C.: Winston.

Rescorla, R. A. 1979. Conditioned inhibition and extinction. In *Mechanisms of Learning and Motivation*, A. Dickinson and R. A. Boakes (eds), pp. 83–110, Hillsdale, N. J.: Lawrence Erlbaum Associates.

Rescorla, R. A. & Cunningham, C. L. 1979. Spatial contiguity facilitates Pavlovian second-order conditioning. *Journal of Experimental Psychology: Animal Behavior Processes*, **5**, 152–61.

Rescorla, R. A. & Furrow, D. R. 1977. Stimulus similarity as a determinant of Pavlovian conditioning. *Journal of Experimental Psychology: Animal Behavior Processes*, **3**, 203–15.

Rescorla, R. A. & Solomon, R. L. 1967. Two-process learning theory: Relationships between Pavlovian conditioning and instrumental learning. *Psychological Review*, **74**, 151–82.

Rescorla, R. A. & Wagner, A. R. 1972. A theory of Pavlovian conditioning: Variations in the effectiveness of reinforcement and nonreinforcement. In *Classical Conditioning II: Current Research and Theory*, A. H. Black and W. F. Prokasy (eds), pp. 64–99. New York: Appleton–Century–Crofts.

Revusky, S. H. 1971. The role of interference in association over delay. In *Animal Memory*, W. K. Honig and P. H. R. James (eds), pp. 155–213, New York: Academic Press.

Revusky, S. H. 1977. Learning as a general process with an emphasis on data from feeding experiments. In *Food Aversion Learning,* N. W. Milgram, L. Krames and T. M. Alloway (eds) New York: Plenum Press.

Rizley, R. C. & Rescorla, R. A. 1972. Associations in second-order conditioning and sensory preconditioning. *Journal of Comparative and Physiological Psychology*, **81**, 1–11.

Rozin, P. & Kalat, J. W. 1972. Learning as a situation-specific adaptation. In *Biological Boundaries of Learning*, M. E. P. Seligman and J. L. Hager (eds), pp. 66–96, Englewood Cliffs, N. J.: Prentice-Hall.

Schwartz, B. 1973. Maintenance of key pecking by response-independent food presentations: The role of the modality of the signal for food. *Journal of the Experimental Analysis of Behavior*, **20**, 17–22.

Shiffrin, R. M. & Schneider, W. 1977. Controlled and automatic human information processing: II. Perceptual learning, automatic attending and a general theory. *Psychological Review*, **84**, 127–90.

Smith, J. C. & Roll D. L. 1967. Trace conditioning with X-rays as the aversive stimulus. *Psychonomic Science*, **9**, 11–12.

Suiter, R. D. & LoLordo, V. M. 1971. Blocking of inhibitory Pavlovian conditioning in the conditioned emotional response procedure. *Journal of Comparative and Physiological Psychology*, **76**, 137–44.

Sutherland, N. S. & Mackintosh, N. J. 1971. *Mechanisms of Animal Discrimination Learning*. London: Academic Press.

Testa, T. J. 1974. Causal relationships and the acquisition of avoidance response. *Psychological Review*, **81**, 491–505.

Testa, T. J. 1975. Effects of similarity of location and temporal intensity pattern of conditioned and unconditioned stimuli on acquisition of conditioned suppression in rats. *Journal of Experimental Psychology: Animal Behavior Processes*, **1**, 114–21.

Trapold, M. A. 1970. Are expectancies based upon different positive reinforcing events discriminably different? *Learning and Motivation*, **1**, 129–40.

Trapold, M. A. & Overmier, J. B. 1972. The second learning process in instrumental learning. In *Classical Conditioning II: Current Research and Theory*, A. H. Black and W. F. Prokasy (eds), pp. 427–52, New York: Appleton–Century–Crofts.

Vandercar, D. H. & Schneiderman, N. 1967. Interstimulus interval functions in different response systems during classical conditioning. *Psychonomic Science*, **9**, 9–10.

Wagner, A. R. 1969a. Stimulus selection and a 'modified continuity theory'. In *The Psychology of Learning and Motivation*, G. H. Bower and J. T. Spence (eds), vol. 3, pp. 1–43. New York: Academic Press.

Wagner, A. R. 1969b. Frustrative nonreward. A variety of punishment? In *Punishment and Aversive Behavior*, B. A. Campbell and R. M. Church (eds), pp. 157–81, New York: Appleton–Century–Crofts.

Wagner, A. R. 1978. Expectancies and the priming of STM. In *Cognitive Processes in Animal Behavior*, S. H. Hulse, H. Fowler and W. K. Honig (eds), pp. 177–210, Hillsdale, N. J.: Lawrence Erlbaum Associates.

Wagner, A. R., Rudy, J. W. & Whitlow, J. W. 1973. Rehearsal in animal conditioning. *Journal of Experimental Psychology Monograph*, **97**, 407–26.

Wasserman, E. A., Franklin, S. R. & Hearst, E. 1974. Pavlovian appetitive contingencies and approach versus withdrawal to conditioned stimuli in pigeons. *Journal of Comparative and Physiological Psychology*, **86**, 616–27.

Weisman, R. G. & Litner, J. S. 1969. Positive conditioned reinforcement of Sidman avoidance behavior in rats. *Journal of Comparative and Physiological Psychology*, **68**, 597–603.

Winograd, T. 1975. Frames representations and the declarative-procedural controversy. In *Representation and Understanding*, D. G. Bobrow and A. Collins (eds), pp. 185–210, New York: Academic Press.

Zimmer-Hart, C. L. & Rescorla, R. A. 1974. Extinction of Pavlovian conditioned inhibition. *Journal of Comparative and Physiological Psychology*, **86**, 837–45.

Index